CHEETAH® Purrrrrrr Publishing

Presents

Preparing the Jamaican Scientist

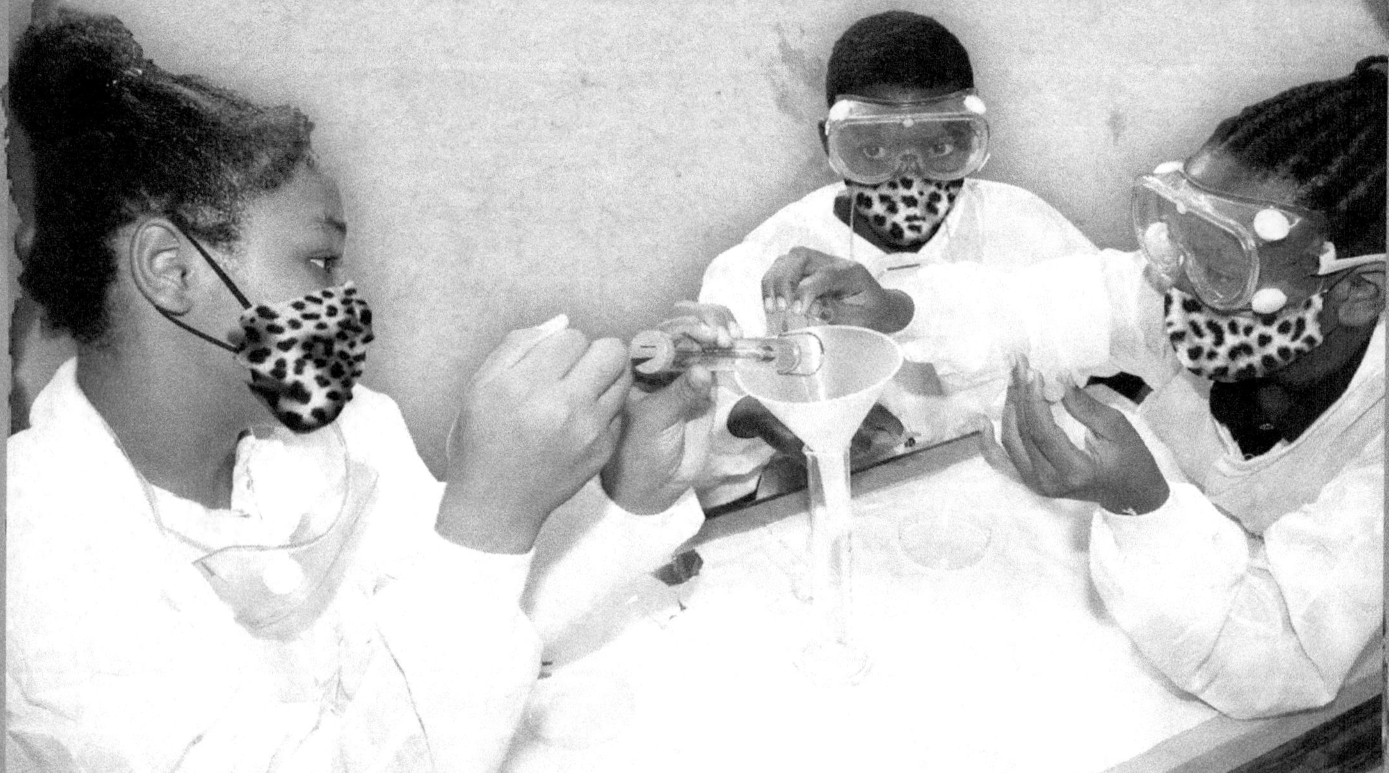

By
Conroy Hall
Paulette Trowers, Juris Doctor

10 9 8 7 6 5 4 3 2 1

Copyright© 2022. All rights reserved.

No part of this book may be reproduced, distributed or transmitted in any form or by any means, including photocopying, recording or other electronic or mechanical methods without prior written permission of the publisher or authors. This book contains proprietary information not yet known to the public.

Although every precaution has been taken to verify the accuracy of the information contained herein, the authors and publisher assume no responsibility for any errors or omissions. No liability is assumed for damages that may result from the use of information contained within.

First published **2022**

Authors: Conroy Hall, Paulette Trowers, JD

Editors and reviewers: Rhoen Kerr, Fiona Porter-Lawson, Pauline Trowers, Steve Gorman, Barbara Jekir, Franklene Frater

CHEETAH™ Purrrrrrr Publishing, an imprint of CHEETAH™ Toys & More, LLC
ISBN-13: 978-1-7328369-2-1
ISBN 10 number: 1-7328369-2-1

Contact information:

CHEETAH Toys & More, LLC
1328 Albany Ave, 2nd Floor
Hartford, CT 06112

Port Antonio P.O.
Portland
Jamaica

www.mycheetahinc.com
WhatsApp or Call: 876-909-6311 / 860-781-1276
info@mycheetahinc.com
paulettetrowers@yahoo.com

Table of contents

Acknowledgements 6

Dear CHEETAH® family 7

Term 1, Unit 1:

The environment and me 8

Chapter 1: The environment 10
   1.1 What is the environment? 12
   1.2 Different environments on Earth 16
   1.3 What is an ecosystem? 28

Chapter 2: Human activities and the environment 38
   2.1 Effects of human activities on the environment 40
   2.2 Pollution 44
   2.3 Solid waste management 47
   2.4 Soil degradation 53
   2.5 Overfishing 60
   2.6 What can humans do to protect the environment? 62
   2.7 Augmented reality 67

Chapter 3: Soil and climate change 69
   3.1 Soil 71
   3.2 Weather and climate 80
   3.3 Climate change 81

Term 1, Unit 2:

Light 92

Chapter 4: Light in our world 94
- 4.1: What is light? 96
- 4.2: Luminous and non-luminous objects 102
- 4.3: Shadows 106
- 4.4: Reflection of light 109
- 4.5: Refraction of light 114

Chapter 5: Sounds in our environment 124
- 5.1: Introduction to sound 126
- 5.2: Properties of sound vibrations 128
- 5.4: Pleasant and unpleasant sounds 139

TERM 2, UNIT 1:

Materials - Properties and uses 146

Chapter 6: Everyday materials 148
- 6.1: Methods for selecting materials 150
- 6.2: Safe use of household items 161
- 6.3: Disposal of household products 168

Chapter 7: Reversible and Irreversible Changes 176
- 7.1: What are reversible and irreversible changes? 178
- 7.2: Comparing changes 189

TERM 2, UNIT 2:

Body systems 196

Chapter 8: Human body systems 198

- 8.1: What is a body system? 200
- 8.2: The human digestive system 202
- 8.3: The musculoskeletal system 205
- 8.4: The excretory system 211

TERM 2, Unit 3:

Mixtures and separation techniques 219

Chapter 9: Mixtures and separation techniques 221

- 9.1: What is a mixture? 223
- 9.2: Classifying mixtures 224
- 9.3: Separating mixtures 228

TERM 3, Unit 1:

Diet and food 239

Chapter 10: Healthy eating habits 241

- 10.1: Balanced diet and how we get to it 243
- 10.2: What is a balanced diet? 243
- 10.3: Consequences of not having a balanced diet 245

Chapter 11: Classification and effects of drugs 255

- 11.1: Drugs and their legal classification 257
- 11.2: Detrimental use of drugs 261

Glossary 269

Acknowledgements

This book is about honour. I give honour to God, who has given me the inspiration, the people, the time, and the resources to make this book possible. I honour the authors, editors, typesetters, photographers and reviewers, among others, who worked tirelessly to move this book from idea to reality.

'A journey of 1,000 miles begins
with a single step.'

-Chinese proverb-

I am Croak, Croaky Croak, that is, and I am a Jamaican scientist. I will motivate, educate and inspire you the CHEETAH way.

Are you ready?

Come wid mi. Mek wi begin.

Dear CHEETAH® family,

Thanks for the opportunity to present our Grade 6 science textbook, *Preparing the Jamaican Scientist*. Our main goals are to educate, inspire and entertain students while they learn the CHEETAH® way.

This textbook aligns fully with the New Standards Curriculum (NSC) and is organised by terms, units and chapters, in the order outlined in the NSC. Additional sections include engaging activities, assessment questions and the latest technology utilising *augmented reality* (AR).

We also have a workbook, *Preparing the Jamaican Scientist PEP Workbook Grade 6*. This contains practice questions, answers and explanations that will enhance learning and prepare students for the national PEP examinations. These science books prepare young minds for both the classroom and life. Say with me, 'I-PEP for life.'

Students may qualify for one of CHEETAH's® national scholarships. Call us or speak with your principal about how you may be eligible. Stay in touch. Your feedback is valuable as we grow and creatively cater to your educational needs.

We want you to journey with Croaky Croak™, a wise Jamaican frog and scientist, who will provide overall guidance and general tips.

We are CHEETAH®, chasing and capturing your dreams with you.

Term 1, Unit 1:
The environment and me

Words worth knowing

The list contains the scientific vocabulary from this unit.

- aquatic ecosystem
- climate
- climate change
- deforestation
- ecosystem
- environment
- environmental conservation
- erosion
- interdependence
- landfill
- microscopic
- overfishing
- overpopulation
- pollution
- reforestation
- slash-and-burn farming
- soil conservation
- soil degradation
- solid waste
- terrestrial ecosystem
- topsoil
- urbanisation
- weather
- wetlands

This is an important note for teachers, parents and students. As you go through the book you will see the adult supervision sign on some activities. These can be dangerous, so we recommend you do them under the supervision of an adult.

ATTENTION

Adult supervision required for this activity.

Without delay, let us find out why is it important to care for the environment.

Chapter 1:
The environment

Chapter objectives

- ✓ Formulate a definition of the environment.
- ✓ Justify the importance of conserving the natural environment.
- ✓ Recognise the need for and importance of conserving living things and the environment to sustain the balance in the ecosystem.
- ✓ Observe, collect and record information regarding the interacting factors within an environment.

CHEETAH Science Fiction

Unlikely Friends

'It's time to go into the deep, Rema,' said Heelie. Heelie was preparing her daughter to attach herself to the underbelly of a shark.

'Oh, mother, must I leave you?' Rema wailed.

'Yes, honey. This is for your survival. This is how you get a free ride across the ocean. You will float along, save your energy and get your food. Plus, you will be protected from bigger fish that want to eat you!' Heelie was trying to cheer Rema up. This was the way of the remora fish. They had made an agreement with the sharks hundreds of years ago. They ate the scraps that fell from the sharks' meals, and in turn, kept the sharks clean by removing pests from their bodies.

'But... but what if he turns around and eats me?' Rema was terrified.

'Well, you will avoid the sandbar and the lemon sharks. They have a chip on their shoulders and can get really aggressive. Avoid sharks that are dark or yellow. Stick with the whales or the white shark.' Then she continued, rehearsing all the information that she felt Rema needed to ensure survival: 'You will also be able to eat the pesky parasites that stick on a shark's skin and become a bother. There will be plenty of food for you, and the shark will love you for helping!'

Rema was in deep thought, beginning to believe her mother's words at last. Her mother continued, 'You can even take parasites from a shark's mouth at times, and he will allow you!'

'Its mouth?' Rema asked in renewed terror.

'Its mouth, and, like I said, he will love it. Besides, it will stave off the dangerous predators that will try to eat you as well. These creatures will run the other way when they see the sharks coming.'

A gentle swish put an end to their conversation. As coincidence would have it, a white shark swam by their rock. Heelie could not believe their luck. Here was a chance for Rema to have her debut with a friendly-looking white shark. 'Come, honey,' she beckoned Rema. 'Quick! This is a perfect moment. You will be fine. Just do whatever you see me do.'

A timid but curious Rema peeped out as the shark swam by their rock again. Heelie slipped into the deep and attempted to fasten her aging dorsal fins to its underside. These fins were like suction cups on the head of all remora fish, adapted for this task. Somehow, her fins held fast. Rema watched, fascinated for a moment, then realised that her mother and the shark were swimming away. 'Mother, wait!' She darted through the water herself.

She reached the underbelly of the shark in a flash, youth, excitement and destiny taking over. She put forth her dainty dorsal fins and dug in. It felt like home and she relaxed and enjoyed the ride, the three gliding effortlessly through the bliss of the deep ocean.

1.1 What is the environment?

Look around you. What do you see? The things you see in your surroundings depend on where you are at that time. Look at Figure 1.1. Do your surroundings have some of the same things in the pictures? Can you give another name for `surroundings'? What type of environment does each picture represent? Can you identify the things you see in each environment?

Figure 1.1: Four different environments.

The word environment comes from the French word *environ*, which means 'our surroundings'. Your *environment* is everything around you. It includes areas where you play, live and go to school.

The environment includes many different factors which are living and non-living things. The *living factors* of the environment include humans, plants, animals, insects, bacteria and any other living things. Living factors are also called *organisms* or *life forms*.

Figure 1.2: Living things in their natural environment.

The *non-living factors* in an environment include the air living things breathe, the water that covers 70% of the Earth's surface, the types of soil plants need to grow, sunlight,

weather and climate. These are just a few. Living and non-living things interact with each other, influencing living conditions and survival on Earth.

There are two types of environments: natural and artificial. The natural environment includes living and non-living things in a space not modified by humans. The artificial environment consists of human changes to the natural environment. While the natural environment includes mountains, rivers, streams and the ocean, the artificial environment includes buildings, roads and bridges.

Justify the importance of conserving the natural environment

Taking care of the environment is an important decision to make. Why is this so? Our natural environment includes the air we breathe, the crops we grow for both food and medicine, other plants and animals and the non-living things we interact with. Without this interaction with other living and non-living things, humans will not survive on Earth.

We, therefore, need to conserve and protect the environment. It is our home. So, we need to put measures in place to prevent permanent damage to our natural surroundings.

Activity 1.1.1: The environment

1. Look around you. Make a list of ten living and ten non-living things in your current environment.

Living factors	Non-living factors

Let us examine how different environmental factors affect living things.

1. Choose FIVE of the living factors you listed in the table. How can these affect you in your environment?

2. Choose FIVE of the non-living factors you listed in the table. How can these affect you in your environment?

3. List FIVE ways in which you can help to preserve the outdoor environment.

4. What factors in your school environment are different from those in your home environment?

5. State TWO ways in which human activities destroy the natural environment.

6. Why is it important to preserve the environment?

Now let us further examine how different environmental factors affect living things.

Activity 1.1.2: Find me

Materials:

- a camera or magazine pictures
- a large piece of cartridge paper or cardboard

Procedure:

1. Collect pictures that look similar to your environment. Use these pictures to make a collage of your surroundings. You can take pictures yourself or find pictures from other sources, such as the internet or a magazine.

2. Label the living and non-living factors in your pictures.

3. Choose ONE living factor from your collage. How would removing this factor affect the other living factors in the environment you created?

4. Choose ONE non-living factor from your collage. How would removing this non-living factor affect the living factors in the environment you created?

5. Give TWO other factors, not shown in the collage that you may find in the environment you created.

6. How is water important to the environment you created?

7. How are air and light important to the environment you created?

Did you know that Jamaica has 73 indigenous species of orchids?
60 Interesting Facts About Jamaica - The Fact File
https://thefactfile.org/jamaica-facts/

1.2 Different environments on Earth

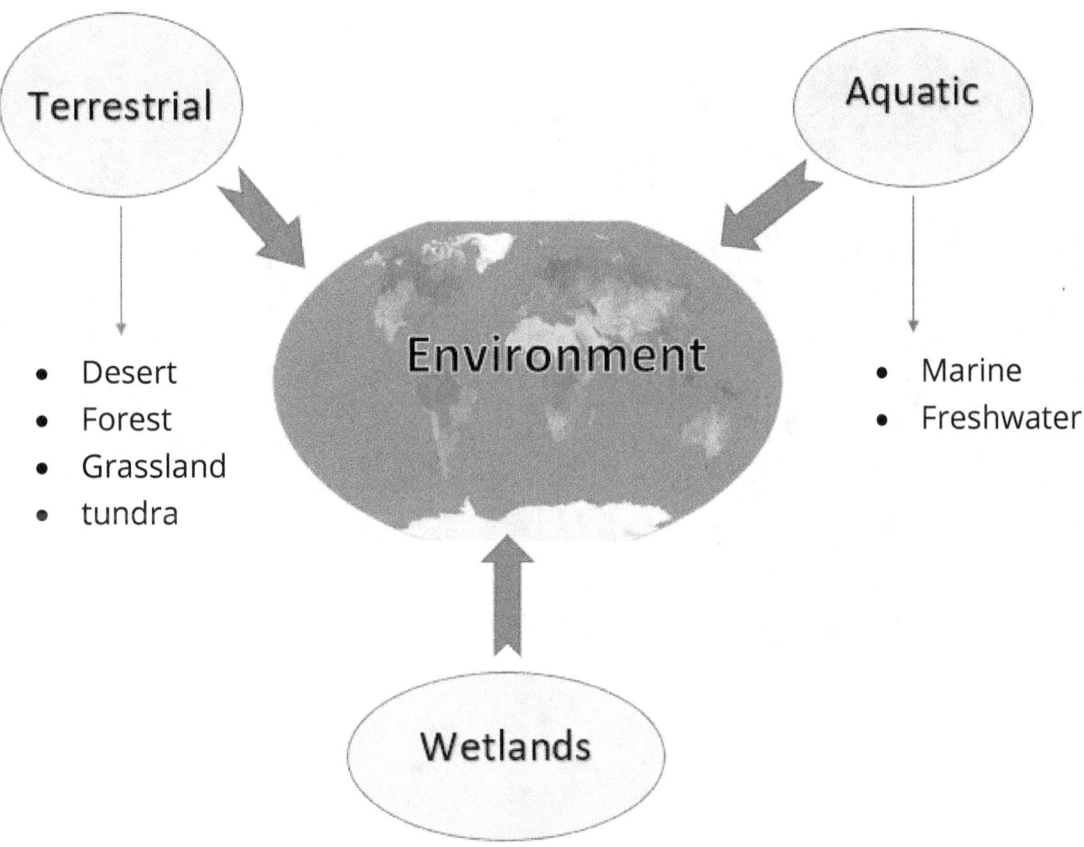

Figure 1.3: Earth's different environments.

Earth has different environments: terrestrial or land environments, aquatic or water environments and wetlands. Scientists further separate these environments by regions that have similar weather, temperature and organisms.

Each environment has unique soil, vegetation, terrain and climate that support the many life forms there.

Protected environments of Jamaica

Natural environments often compete and overlap with artificial environments built by humans, such as factories, cities, farms and roads. Some human activities are destroying the environment. To protect the environment, countries sometimes establish protected areas. Persons cannot develop, encroach on or damage protected areas. The Jamaican government has marked some areas as protected. See examples of protected areas in Figures 1.4 to 1.6.

Figure 1.4: *Manatee Bay is part of Portland Bight Protected Area in St. Catherine and Clarendon.*

Figure 1.5: *The protected Blue and John Crow Mountains in Jamaica.*

The Montego Bay Marine Park is a protected area of seawater on Jamaica's north coast. This Park is managed by a Trust set up to protect the area. The Trust also educates the surrounding community about the importance of the natural environment and the resources it contains.

Figure 1.6: *The protected Montego Bay Marine Park, Jamaica.*

Figure 1.7: *The Negril Watershed Environmental Protection Area, Jamaica.*

Terrestrial environments

1. Deserts

Deserts are one of the driest environments, only receiving about 30 centimetres of rainfall per year. Desert temperatures can range from 10°C in winter to 49°C in summer, with hot days and cool nights. Deserts can be found on every continent and cover one-fifth of the Earth's land area.

Figure 1.8: *Desert environment.*

We often think of a desert as a sandy place without many living factors. However, the desert has many organisms, including plants like cacti and agave, as well as animals such as coyotes, javelinas, lizards and snakes. The cactus is a special type of desert plant that stores water in its stem and uses its spiny leaves for protection. Many desert animals stay in underground tunnels or shady spots during the day and only come out at night to search for food.

2. Forests

2a. Tropical rainforests

Tropical rainforests are forests that are warm and wet, receiving an average of 300 centimetres of rainfall per year. Tropical rainforests are found near the Equator, which means the amount of sunlight they receive does not change very much from season to season. Tropical rainforests have a constant temperature that ranges from 20 to 30°C year-round, and contribute approximately 20% of the oxygen on Earth. The Amazon Rainforest in Brazil is one example of a tropical rainforest.

Figure 1.9: *A jaguar in a tropical rainforest.*

Plants in tropical rainforests have broad leaves and form a dense canopy. Trees can be as tall as 35 metres (taller than a ten-storey building). Tall trees form such a thick canopy that only a small amount of sunlight can reach the forest floor.

Figure 1.10: *A macaw.*

Most of the animals in tropical rainforests live in trees. Tropical rainforests also contain many unique animals found nowhere else on Earth. This is the only ecosystem where you can see organisms like poison dart frogs, jaguars and macaws.

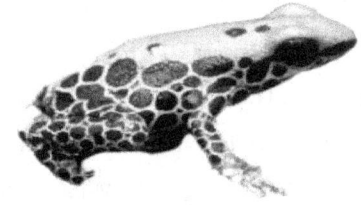

Figure 1.11: *Poison dart frog.*

2b. Temperate forests

Temperate forests are found in places that have four seasons: winter, spring, summer and autumn or fall. Many temperate forests contain a variety of deciduous trees, such as aspen, oak or maple. These kinds of trees lose their leaves in the autumn and regrow them in the spring.

Figure 1.12: *Temperate rainforest.*

Figure 1.13: *Rainforest ferns.*

Animals such as frogs, turtles, hawks, owls, raccoons, porcupines and foxes all live in temperate forests. Many varieties of plants such as moss, wildflowers, ferns and shrubs can also grow and thrive in this ecosystem.

3. Grasslands

Grassland ecosystems have many grasses or flowers, but very few trees. In some places, there are no trees at all. They receive more rainfall than deserts, but less than forests.

Tropical grasslands are often referred to as *savannas*. Savannas have a dry season and a rainy season and receive just enough rain to allow grass, small shrubs and a few large trees to grow. Tall grasses grow with long roots to allow plants to find water in the dry, hot summer. Many different animals, such as lions, elephants, giraffes and zebras, live in this unique environment.

Figure 1.14: *Grassland.*

Figure 1.15: *Elephants on the savanna.*

Temperate grasslands are found in colder regions of North America, Russia and South America. They are also called *prairies* or *plains* and are covered with grass but very few trees. These kinds of grasslands are often used for growing crops because of their nutrient-rich soil. Animals that live in temperate grasslands include prairie dogs, bison and snakes.

Figure 1.16: *Snake in temperate grassland.*

4. Tundras (polar regions)

The *tundra* is a cold and dry environment that has very long and cold winters. Temperatures in the tundra can reach -30°C and it snows a lot! In the winter, the days are short, and several days of the year are completely dark.

Figure 1.17: *Penguin and polar bear in their natural tundra environment.*

In the summer, the days are long (as many as 24 hours), and on these days, the sun shines throughout the night. On these days, the sun can be seen just over the horizon as though it were sunset. Much of the snow melts in the summer, but a special layer of the soil stays frozen all year.

Since the tundra is so cold and dry, not many plants and animals live there. But there are some plants and animals that can survive this tough environment. In some tundra ecosystems, trees grow alongside small plants like lichens, mosses and wildflowers. Animals such as polar bears, caribou, lemmings and arctic foxes also live in tundra environments.

5. Wetlands

Wetland environments are neither aquatic nor terrestrial. They are lowland areas in which the land is covered by water seasonally or permanently. Wetlands have both aquatic and terrestrial species and can be natural or man-made. Some wetlands are temporary, lasting for a few days, weeks or months—seasonally, while others are permanent. The water in a wetland can be fresh, salt or in between and can be flowing or stationary.

Figure 1.18: *Marine mangrove.*

Table 1.1: Jamaican wetlands

	Wetland	Location
1	Great Morass	St. Thomas
2	Black River Morass	St. Elizabeth
3	Great Salt Pond	St. Catherine
4	Salt River Swamp	Clarendon
5	Canoe Alley	Manchester
6	Hague Swamp	Trelawny
7	Negril Swamp	Westmoreland and Hanover

Figure 1.19: Map showing some locations of wetlands in Jamaica.

Aquatic environments

1. Freshwater environment

Rivers, ponds and lakes are all examples of *freshwater environments*. Rivers are large, natural streams of flowing water found in many locations around the Earth. Some examples of rivers in Jamaica are the Black River in St. Elizabeth, the Great River in St. James and the Rio Grande in Portland. Unlike rivers, lakes and ponds are basins of water that are surrounded by land and can be natural or man-made.

Figure 1.20: *Freshwater environment—river.*

Many organisms, such as turtles, freshwater fish, insects, algae and water lilies, live in freshwater environments.

Figure 1.21: *A freshwater environment – pond.*

2. Marine ecosystems

Marine ecosystems are the largest natural environment. They include salt marshes, mangrove swamps, seas and oceans. Marine ecosystems are found all over the world and occupy more than 70% of the Earth's surface. A variety of organisms live in marine ecosystems, including animals like clams, crabs, fish and coral.

Figure 1.22: *Seagull eating sea crab.*

Figure 1.23: *Flourishing saltwater environment.*

Did you know that Jamaica currently has four sites designated as wetlands of international importance?
Learn more about wetlands in Jamaica at:
https://www.ramsar.org/wetland/jamaica.

Learn how you can participate in the next `**World Wetland Day**' held on the 2nd of February each year.
World Wetlands Day | Convention on Wetlands (ramsar.org)

Activity 1.2.1 Environments on Earth

Let's look at the environment on Earth. Why would a frog in a pond stay close to a flower? Recommend a rule or law that would protect wetlands.

Select the ecosystem that is most likely a wetland.

Circle the animal that can most likely be found in a tundra environment.

As human communities expand, they affect the environment of other living things. Choose ONE terrestrial environment.

    a) Give FIVE ways human activity affects this environment.

    b) How does damaging this environment affect humans?

    c) Why does protecting this environment matter?

Activity 1.2.2: Terrestrial environment in Jamaica

Materials:
- camera/internet photos
- glue stick
- paper
- play dough

Procedure:

1. Make a scrapbook showing some of the different terrestrial environments in Jamaica. These can include forests, wetlands, grasslands and desert-like areas. For each environment, research information about the environment, such as its terrain, temperature, plants and animals. In your scrapbook, include pictures of the plants, animals and other environmental factors that you have researched for each type of environment.

2. Use play dough to create a 3D map of Jamaica. On your map, mark the location of three specific terrestrial environments, for example, Blue and John Crow Mountains National Park.

Now let's look at the terrestrial environment.

3. In the table below, list the living and non-living things found in the terrestrial environment.

Name of environment	Living things	Non-living things
1.		
2.		
3.		

4. Did any of your environments have any living or non-living things in common? Give TWO examples.
5. Look back at your list. Were any living or non-living things unique to only one environment? Give TWO examples.

1.3 What is an ecosystem?

Ecosystems and the environment

In all environments, living things and non-living things work together to make life possible. All the factors of an environment that work together make up an ecosystem. An *ecosystem* is a community of organisms that live with and depend on each other, as well as on non-living materials. Some non-living things an ecosystem needs are sunlight, air, water and land. An ecosystem needs living and non-living factors to survive.

How ecosystems work

Let's look closer at how an ecosystem works by examining a pond. A pond is an example of a shallow, freshwater ecosystem. Animals, like fish, frogs, ducks and water bugs, can make up the living factors of a pond. Algae, bacteria and plants are other living things found in a pond. The excreted waste from the animals provides nutrients for the plants. Bugs that visit the flower form food for the frogs.

Figure 1.24: *Frog in a pond.*

Some non-living factors affecting the pond are water, chemicals or minerals in the water, oxygen in the water, soil at the bottom of the pond, sunlight and water temperature.

The organisms that live in an environment need the right kinds of non-living factors to survive. These non-living things make the environment healthy. For example, if there is not enough oxygen in the pond water, some species of fish living there will die.

Figure 1.25: *Whistling ducks in their environment.*

What is a stable ecosystem?

A *stable ecosystem* explains how living and non-living things work together in an ecosystem to provide balance. An ecosystem is considered stable when it does not experience any big or sudden changes over a long period. An ecosystem is also considered stable if it can return to normal (balance itself) after environmental changes.

What factors affect the stability of an ecosystem?

Many factors can make an ecosystem stable or unstable, such as changes in the climate, the number of prey or predators or human activity. Unstable ecosystems may experience major changes over short or long periods and can cause a serious imbalance. This is demonstrated through conditions such as drought, loss of food sources or habitat and activities such as overuse of chemicals. An unstable ecosystem may cause living things to die in large numbers.

Figure 1.26: Overgrazing can cause instability. The photo shows a flock of sheep grazing on a small area of land.

Figure 1.27: Deforested area. Deforestation destroys the habitats and food sources of living things and can cause instability in the ecosystem. Noise from machinery also frightens away animals from their habitats.

Test yourself!

1. Name ONE ecosystem and explain how the ecosystem works.
2. Give ONE example of changes to the ecosystem and how they affect the living things in the environment.

Interdependence of organisms in the environment

In any environment, living things rely on each other to grow and survive. This is called *interdependence*.

Here are some examples of interdependence.

- Animals depend on plants for food; both goats and cows eat grass.
- The droppings from animals form nutrients for plants; the plants then grow and become food for the animals.
- **interdependence.**
- The animals grow and become food for other animals.
- When animals die, they rot (decompose) and their nutrients go back to the soil for plants to absorb.

Examples of interdependence

1) One example of interdependence is the relationship between the shark and the remora fish. The remora fish cleans the shark and, in turn, receives food scraps left over from the shark's hunt.
2) Another example of interdependence is the role of the algae eater, e.g. suckermouth catfish in a fish tank. The catfish eats the algae that grow on waste the fish give off. As it eats, it cleans the tank, stops waste buildup and balances the tank environment.
3) The relationship between the cow and the egret is another example of interdependence. The egret stands on the cow and eats flies, ticks and other insects from the cow's back. Some of these small animals, such as ticks, harm the cow. The egret also acts as an alarm, alerting the cow to predators.

CHEETAH® PREPARING THE JAMAICAN SCIENTIST

Did you know that Jamaica has 28 species of birds found only in Jamaica? This is part of Jamaica's rich diversity of flora and fauna.

28 Species of Birds Unique to Jamaica - WorldAtlas

Before you go the next page, are you sure you are ready to do so? Are there any words on this page you don't understand—like species, diversity, flora, or fauna?
I always keep a dictionary nearby or use my phone to look up new words.
What do you do when you come across a word you don't know?

31

Activity 1.3.1 Interdependence

Now, state TWO examples of interdependence between the animal and plant in the picture. Use this information to write a title for the picture.

Figure 1.28: *brown and white bird on red and green plant photo*

CHEETAH™ collaboration corner

Activity 1.3.2: What is interdependency within the environment?

In this group activity, you will examine how various non-living and living factors impact an imaginary environment.

1. Using the list, create cards with each item so that each child in the class has a card.
   a. air
   b. sunlight
   c. soil
   d. worm
   e. bird
   f. cat
   g. water
   h. flower
   i. lizard
   j. bee
   k. stone
   l. plant

2. Two students are to be selected to explain the relationship between the two factors noted on their cards. For example, if one student has a card marked 'air' and another a card marked 'cat,' then the student with the card marked 'air' can state, 'Cats need air to breathe.' The student with the card marked 'cat' can say, 'All living things need air to breathe.'

3. Each person in turn explores the relationships between their factor and the other factors in the class (inter-relationships).

4. After the activity has ended, students will work together in small groups to write a summary of one interrelationship with each factor.

Now, use your summary to look at interdependency.

5. What were the living and non-living factors found in your environment?
6. If all the non-living factors were removed, what would happen to your environment?
7. If all the living factors in your environment were removed, what would happen?
8. Identify an environment and describe the inter-relationship between two living things within that environment.
9. Identify an environment and describe the inter-relationship between a living and a non-living factor present in the environment.

For more practice questions use the *CHEETAH® PEP Science Practice Questions workbook*. It has over 400 questions to sharpen your memory.

Activity 1.3.3: Information and communication technologies (ICT) and me: Sustainable development

Use the websites listed below to research the topic 'sustainable development'.

- https://www.who.int/health-topics/sustainable-development
- What Is Sustainable Development? - Monash Sustainable Development Institute

Prepare an electronic brochure to highlight ways in which you can be a part of sustainable development in your community.

Use an application (app), such as *Animoto* and add music or sounds to the pictures to create a video entitled, 'Sustainable Development - Protect the future of the environment.'

Activity 1.3.4: Word find

In this activity, you will find the keywords used in Chapter 1. Find and circle the words in the puzzle.

- environment
- ecosystem
- conservation
- adaptation
- physical features
- preservation
- interdependence
- mangrove swamps
- coral reefs
- terrestrial
- aquatic
- wetlands
- tundra

E	N	V	I	R	O	N	M	E	N	T	C	H	A	N	G	E
N	C	M	A	N	G	R	O	V	E	S	W	A	M	P	S	G
V	I	N	E	R	Y	U	I	O	P	A	S	D	F	G	A	N
I	T	N	E	C	O	S	Y	T	E	M	Z	X	C	A	Q	C
R	A	D	A	D	A	P	T	A	T	I	O	N	A	S	T	O
O	U	V	B	N	N	M	Q	W	E	N	R	Y	U	I	O	R
N	Q	I	N	F	A	E	T	U	R	E	S	R	E	E	F	A
M	R	O	C	K	V	I	P	M	A	R	I	N	E	S	O	L
E	N	O	I	T	A	V	R	E	S	N	O	C	E	S	T	R
N	C	I	N	O	S	S	A	N	D	E	S	E	R	T	S	E
T	U	N	D	R	A	P	E	N	E	R	S	I	O	N	C	E
G	R	A	S	S	L	A	I	R	T	S	E	R	R	E	T	F
I	P	H	Y	S	I	C	A	L	F	E	A	T	U	R	E	S
P	R	E	S	E	R	V	A	T	I	O	N	S	N	U	S	J
W	E	T	L	A	N	D	S	F	R	O	G	S	O	I	V	E

Evaluate yourself!

Use this evaluation grid to check your understanding of the concepts discussed in this chapter. Read each statement below and insert the symbol that best shows how well you feel you understand the concept. Ask a teacher or parent to help you go over any areas that are still unclear, or that you do not feel you have mastered. Be honest!

I got it!

I need to do more work.

I do not get it. I need help.

In this chapter:

	I got it!	I need to do more work.	I do not get it. I need help.
1. I can explain the meaning of the word environment.			
2. I know why it is important to take care of the natural environment.			
3. I understand the need for and importance of caring for living things and the environment.			
4. I can observe, collect and record information about the environment.			

CHEETAH® PREPARING THE JAMAICAN SCIENTIST

According to Thomas Edison, 'Our greatest weakness lies in giving up. The most certain way to succeed is always to try just one more time.'

What do you think?

Mek wi hop to di nex chapta.

Let's go! Mek wi leap fi knowledge!

Chapter 2:
Human activities and the environment

Chapter objectives

- ✓ Outline the effects of human activities on the environment
  - deforestation
  - poor agricultural practices
  - urbanisation and overpopulation
  - pollution
  - solid waste management
  - overfishing
  - mining.
- ✓ Show concern for the impact of humans on the environment.
- ✓ Propose measures to reduce/eliminate selected sources of solid waste pollution.
- ✓ Describe the factors that cause soil degradation.
- ✓ Explain how soil degradation can be prevented.
- ✓ Be aware of their responsibility to carry out good environmental practices.

CHEETAH Science Fiction

The Schoolyard

'Take your seats. I have something pretty exciting for you today!' Miss Trowers announced. The children were murmuring, hesitant to believe anything a teacher called exciting would be fun. But to everyone's surprise, Miss Trowers reached into her bag and pulled out a black device, holding it up for the class to see.

'I can see the confusion on your faces, so I won't keep you guessing. My friend Josh works for the National Planning and Environmental Agency (NEPA), and he found a way for us to get some old carbon dioxide testing metres to try out.'

It wasn't exactly an expedition to the top of Blue Mountain, but it was better than taking notes all day. 'Before I hand these out, we need to take some notes.'

A groan filled the classroom as Miss Trowers turned to the board and started scribbling numbers. 'A normal range of carbon dioxide found outside should be somewhere between 300 PPM and 400 PPM. PPM is parts per million. When it starts to get higher than that, it's a bad sign. Just like the smoke alarm would go off in your house, we can use these devices to see if the carbon dioxide is too high outside.'

After a quick tutorial, the kids were paired up and marched outside to start measuring the levels of carbon dioxide around the schoolyard. Normal readings were being marked down in notebooks left and right. One team approached a corner of the field, usually avoided because of its bad odour, and called out to the teacher.

'Umm, Miss Trowers. I think ours is broken!' Ben yelled. 'Look! It says 800 PPM. That seems bad!'

When Miss Trowers called Josh later that afternoon, he admitted that it was high. In fact, that the reading was nearly toxic. He assured her a NEPA team would investigate the matter. The NEPA team arrived the following day and used their standard metres. To their surprise, the reading was extremely high. They then called in a team to dig up the spot and found that the school had been built on an old landfill.

The children were shocked at what they saw. Newspaper, plastic and glass bottles, diapers, old clothing, car tyres and plastic bags, none of which had decomposed. The NEPA recommended that the area be sealed off.

The children were pensive for the rest of the day. Those items in the old landfill had been there for decades. They were being affected by the actions of people they had never even met. How would their actions affect those who would come after them? How are their actions affecting the future of their planet?

Activity 2.0: Human impact on Earth's environment

Before you read this chapter, complete this group activity.

Form a group of THREE and discuss how you think human beings affect the natural environment and how our activities affect how much carbon dioxide is in the air around us. Make a list of at least FIVE important ways people affect nature.

2.1 Effects of human activities on the environment

As we go about our daily activities, we change or impact our environment. Some of these changes are bad and can damage the environment.

The following are some human activities that change the environment.

Deforestation

Deforestation or logging is the cutting or burning down of trees in large numbers over a wide area of land. People remove trees for firewood, make paper, build structures, make charcoal, farm and mine for gold, bauxite and marl.

Deforestation harms the environment. For instance, removing large numbers of trees:

Figure 2.1: Deforestation.

- ✓ causes the bare land to be exposed to water, sunlight and wind.
- ✓ destroys the habitats of some insects and animals.
- ✓ causes carbon dioxide to build up in the air. Plants remove carbon dioxide from the air to make their food (photosynthesis). Carbon dioxide traps heat in the environment, which increases the temperature of the Earth's surface. Trees help by absorbing carbon dioxide and releasing oxygen in exchange.

- ✓ increases water erosion of the soil and causes flooding in some areas because the plant roots that hold the soil in place are removed. The land can become dry, leading to desertification (the drying out of the soil), which makes it unable to support plant growth.
- ✓ decreases rainfall because there are no plants to give off water to the air.

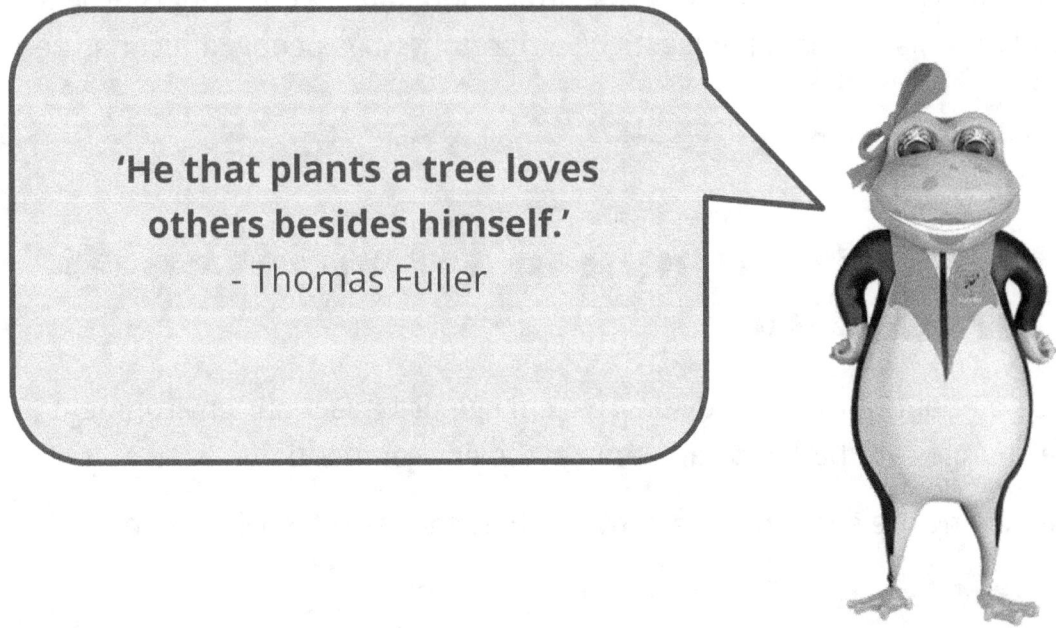

'He that plants a tree loves others besides himself.'
- Thomas Fuller

Poor agricultural practices

Farmers who want to grow food crops must clear the land and prepare the soil. However, some methods damage the soil. Let us look at some of these methods.

A. Slash-and-burn farming

Have you seen people burning bushes on the land to prepare it for planting crops? This method is called **slash-and-burn farming**. Farmers use the slash-and-burn method to cut down and burn plants to create new fields for planting crops and rearing cattle. This slash-and-burn farming practice affects the environment by causing:

- ✓ deforestation
- ✓ habitat loss for many animals
- ✓ increased soil erosion and landslides
- ✓ loss of water from the soil
- ✓ the death of organisms in the soil which are needed to recycle nutrients.

Figure 2.2: Land burning.

Figure 2.3: After the land is burnt.

B. Overuse of chemicals in the soil

Farmers often use chemicals to get rid of insects and other animals that would eat their crops. However, the **overuse of chemicals** in the soil destroys nutrients or organic matter in the soil. Overusing chemicals like pesticides, weedicides, herbicides, fungicides and fertilisers used in farming can increase toxins that poison the soil.

Figure 2.4: Overuse of fertiliser.

Figure 2.5: Overuse of pesticide.

Urbanisation and overpopulation

Another cause of environmental change is **urbanisation**. **Urbanisation** occurs when people move in large numbers over time from rural areas (the countryside) to urban (town) areas. For example, when people move from rural areas to urban areas to find better jobs, health care or schools.

Examples of urban communities in Jamaica are sections of Downtown Kingston,

Spanish Town, Portmore, Montego Bay, Ocho Rios and Mandeville.

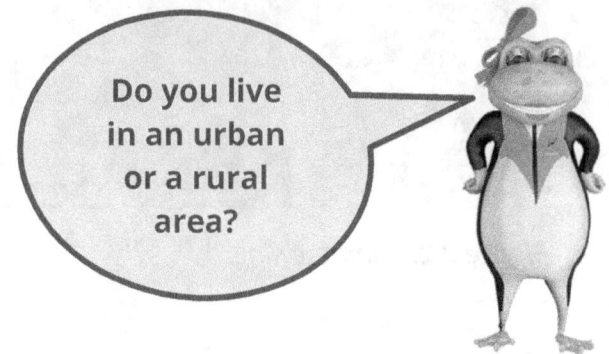

Figure 2.6: City (urban area).

When people move into these cities, the areas become crowded and overpopulated. You will therefore find large numbers of people working, living and occupying a small space.

The **number of people** then becomes too large for the living space and the available food, water and shelter. This clustering of people and lack of resources, such as affordable housing and unemployment, can cause several people to become homeless.

Figure 2.7: Street (homeless) person.

People have a lower quality of life when there is hunger, poverty and lack of proper housing.

High-rise buildings

Overcrowding

Figure 2.8: Results of overpopulation.

CHEETAH® PREPARING THE JAMAICAN SCIENTIST

Did you know the average person generates 4 pounds of trash each day?

2.2 Pollution

Have you ever dropped a candy wrapper outside the bus window or tossed a plastic bottle in the garbage with the rest of the trash? Can you imagine millions of people doing the same thing?

Pollution is adding anything to the environment that contaminates the environment and harms living things. Three main types of pollution are air pollution, water pollution and land pollution.

Figure 2.9: *Air pollution (left), land pollution (centre) and water pollution (right).*

Air pollution

Air pollution occurs when humans add waste and other substances that are harmful to the air and living things. **Air pollution** contributes to different illnesses, such as heart disease and asthma.

Figure 2.10: Smog (smoke and fog) covered city.

Burning fossil fuels (gasoline, coal and cooking gas) for energy release poisonous gases into the air, making it unhealthy for living things. Fossil fuels are used to power electrical equipment, automobiles and many other technological devices.

Figure 2.11: Gases given off from motor vehicle exhaust are poisonous.

Water pollution

Water pollution occurs when waste (garbage) is disposed of in the water. The main cause of water pollution is poor garbage disposal. The three types of waste that cause water pollution are:

1. chemical waste from factories
2. untreated or partially treated sewage
3. rotting household garbage

How do these impact the water? They give the water a bad odour and colour, making the water unsuitable to drink, wash, cook or bathe. Polluted water also causes water-borne diseases, such as malaria and cholera in humans. Polluted water doesn't just make humans sick; it also affects the animals living in the water and causes diseases.

Figure 2.12: *Water pollution with garbage (left) and pollution caused by sewage (right).*

Land pollution

Land pollution is the disposal of waste on the land. It is the dumping of garbage or other pollutants such as factory waste and waste oil on the soil or in gullies. Poor management of landfills also causes pollution. Landfills are places that collect and burn garbage. Mining for rocks and minerals, or spraying pesticides on farmlands, also releases harmful substances into the air and pollutes the land. Like air pollution, land pollution harms soil organisms, poisons plants and causes diseases in humans.

Figure 2.13: *Water pollution (left) and pollution in landfills (right).*

Preventing pollution

Humans can prevent pollution in the following ways:

- creating laws to manage waste and punish offenders
- reducing how much fossil fuel we use
- practising the 3 R's: reduce, reuse, recycle
- using compost and organic manure instead of chemicals and pesticides
- using public transportation, carpooling and walking or cycling more often
- getting energy from sunlight, water and wind.

Solid waste management

Poor solid waste management is another example of our harmful effect on the environment.

2.3 Solid waste management

Solid waste or garbage is unwanted material, including paper, plastic, metal and wood. Not all solid waste material is harmful and not all solid waste is in a 'solid' state. Some solid wastes are liquids such as oil, paint or sewage. When solid waste pollutes water, air or land, it is called **solid waste pollution**.

Figure 2.14: *Garbage truck.*

Image taken from https://www.localgovjamaica.gov.jm/ and adapted for educational purposes.

Solid waste management is the handling of solid waste so it does not harm the environment. For example, recycling bottles and metals, reusing plastic containers, and reducing the use of plastic are good practices. In Jamaica, the government banned single-use plastic bags or what is commonly known as 'scandal bags'.

Figure 2.15: *Garbage truck at a landfill.*

Humans can get rid of solid waste in different ways. When garbage is disposed of in landfills, solid waste is buried far away from humans.

However, not all solid waste is disposed of properly. In numerous communities, people burn solid waste such as plastic, sponge, metal, glass, ceramics, rubber, Styrofoam and PVC pipes. These release poisonous gases into the atmosphere. In some urban communities, people dump garbage along the roadside, in gullies or in their backyards. The odour from the garbage causes air pollution, while the garbage itself accumulates bacteria. This can cause health problems.

Disposing of batteries incorrectly is another cause of pollution. Batteries contain heavy metals such as lead, mercury, arsenic, cadmium and iron that can end up in soil and water. Children who play in soil or swim in water with heavy metals can develop skin rashes, cancers, breathing problems or other illnesses.

Figure 2.16: *Batteries that have been incorrectly disposed of.*

Solid waste that is dumped into rivers, lakes or seas can harm wildlife. Animals mistake garbage for food. Waste often tangles their feet and heads and clogs their stomachs, causing them to die.

Figure 2.17: Animals entangled in solid waste.

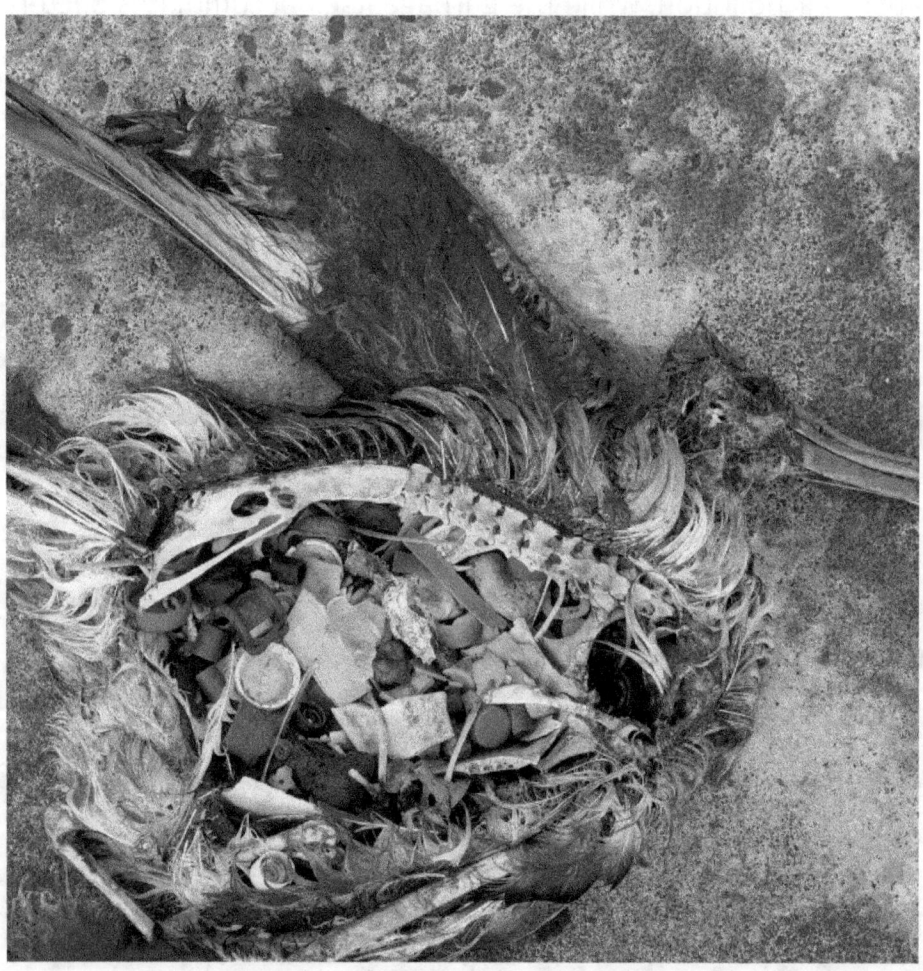

Figure 2.18: Dead bird with solid waste in its stomach.

Improving solid waste management

Humans can improve solid waste management in several ways.

Can you organise a litter clean-up in your neighbourhood?

CHEETAH® PREPARING THE JAMAICAN SCIENTIST

CHEETAH™ collaboration corner

Activity 2.3.1 Solid waste audit

In a group of five, work together to investigate people's attitudes toward solid waste management in your school and community.

Complete the following for your household.

a. How many people live in your home?

b. Copy and complete the table below. State approximately how many of the listed items are brought into your household weekly and how these items are disposed of.

Item	Quantity	Disposal method(s)
plastic shopping bags		
glass bottles		
metal cans		
cardboard boxes		
paper bags		
aerosol cans		
newspapers		

c. Does your family compost, reuse or recycle solid waste? If yes, what types of items?

d. How do you dispose of mobile (cell) phone batteries in your home?

Next

e. Ask FIVE other households to complete the table above.

Later—Group work

f. Form groups of FIVE to analyse the data collected. Create tables and charts to display the data. Discuss and record conclusions made from your data.

g. Share your findings with the rest of the class.

Then—Environmental project

h. Contact NEPA (National Environmental Protection Agency) and JET (Jamaica Environment Trust) about starting a plastic bottle collection drive at school. Then, sell the bottles to buy seeds and tools for a school garden.

How much of your waste do you recycle?

Mining

Mining is the digging of the earth in search of minerals. It removes the topsoil and exposes the subsoil. Consequently, the water in the soil dries out. The soil then becomes unsuitable for growing plants.

Figure 2.19: Large mining operation.

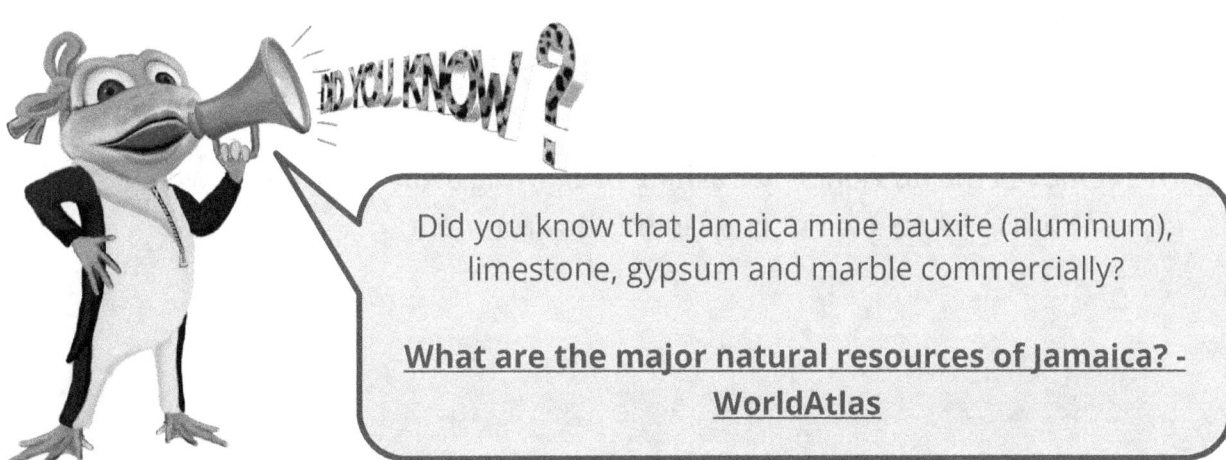

Did you know that Jamaica mine bauxite (aluminum), limestone, gypsum and marble commercially?

What are the major natural resources of Jamaica? - WorldAtlas

2.4 Soil degradation

Grab a handful of soil or dirt or earth. Do you know how important this is? Soil has nutrients that allow us to grow plants for food. It is home to different animals, helps to filter and clean our water and reduces flooding. Anything we do to prevent the soil from doing these things is soil degradation.

What is soil?

Soil is made up of living and non-living matter. The living part of soil consists of organisms such as earthworms, cockroaches, centipedes, millipedes, worms and bugs. The non-living part includes broken-down rocks and decayed remains of animals and plants, which form humus.

Figure 2.20: *A shovel is used to dig the soil.*

Soil degradation

Soil degradation occurs when parts of the soil change and plants cannot get nutrients from it. Soil degradation may be caused by human activities or by the processes of nature.

Causes of soil degradation

- soil erosion
- soil exhaustion
- overgrazing
- mining
- deforestation

Soil erosion

This removes the nutrients by washing or blowing away the topsoil, which can expose rocks. Plants cannot grow in this environment.

Activity 2.4.1 Investigating erosion

Experiment 1: Wind erosion

Materials:

- sand
- small stones
- clay or garden soil
- straw
- 1 small cup of water
- small table

Procedure

1. Working in pairs, set up the following materials on a table, as shown. Materials should be spaced out (not close to each other), and each pile should be approximately the same height.
   - 4 teaspoons of dry sand (A)
   - 4 teaspoons of moist sand (B)
   - 4 teaspoons of dry clay (C)
   - 4 teaspoons of moist clay (D)
   - 6 pebbles (small stones) (E)

2. Using a drinking straw, blow across each pile with the tip of the straw approximately 10 cm from the pile. Use the same amount of effort each time you blow on a pile.
3. Record and discuss your observations.

Complete the table below with your observations from Experiment 1.

Type	Effect of blowing
dry sand	
wet sand	
dry clay or soil	
wet clay or soil	
rock (pebbles)	

Use the data collected during Experiment 1 to answer the following questions:

1. Of the five samples, which ONE was eroded most by the wind?
2. Of the five samples, which ONE was eroded least by the wind?
3. What have you learned from this experiment about soil erosion in real life?
4. How can you use what you have learnt to reduce soil erosion in your community?

Activity 2.4.2 Wind erosion

Experiment 2: Water erosion

Materials:

- two of any of the following containers (small shoe box, small plastic bottle cut horizontally, small Styrofoam food box or baking tray)
- soil taken from one place and mixed
- plant (grass)
- two 500 ml plastic bottles filled with water
- one push pin
- two 100 ml measuring cylinders
- two similar inclined planes
- stop clock

Procedure:

1. Use two of the containers.
2. Put an equal amount of soil in each container.
3. Plant grass taken from a lawn on the soil of one of the containers.
4. Leave the other container with bare, exposed soil.
5. Leave the two containers outdoors for one week.
6. After one week, take each container indoors and set it against an incline (which may just be a piece of board raised on one end by a book).

7. Invert the 500 ml water bottles and punch holes in the bottom using a push pin.
8. Hold the 500 ml plastic bottles 5 cm above each inclined container and quickly pour the 500 ml of water to create 'rain'.

Observation:

Record and discuss the nature of the run-off. Examine the colour of the water and the number of sediments after five minutes, for Experiments 1 and 2.

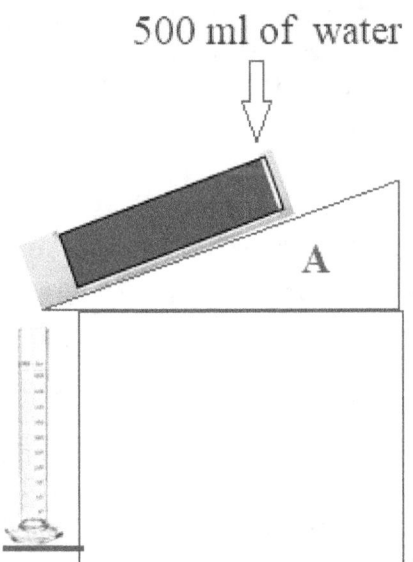
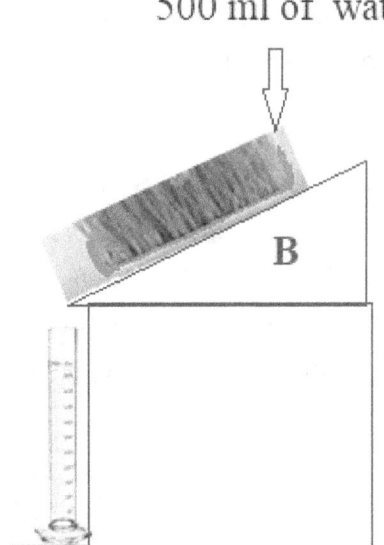

Water erosion through soil (A) and grass (B) on an inclined plane

Use the data collected during Experiment 2 to answer the following questions:

1. In the water erosion activity, which sample, A or B, gave the most run-off?

2. In the water erosion activity, which sample gave more sediments in the water run-off?

3. What conclusions can you draw about water erosion?

4. What have you learned from these models about what happens with soil erosion in real life?

5. What conservation method does the experiment describe?

Soil exhaustion

Soil exhaustion is another type of degradation and is caused by growing the same crops on the same plot of land over a long time. This decreases the nutrients in the soil. After several years, the land produces smaller and fewer crops. When this happens, farmers need to nourish the land. To do so, they either use fertilisers, plant legumes (peas, beans and corn) or leave the land to rest.

Figure 2.21: Soil exhaustion

Overgrazing

Overgrazing also causes soil degradation. Overgrazing occurs when there are too many plant-eating animals on a small plot of land. Animals like cows, horses, sheep and goats eat plants right down to the ground. When overgrazing exposes the soil, wind and rain cause soil erosion. To prevent overgrazing, farmers usually rotate the field these animals graze on to allow the plants to recover.

Soil conservation practices

To help protect against soil degradation, many farmers practise soil conservation. **Soil conservation** takes steps to prevent soil degradation.

Some of these steps are:
- reforestation
- organic farming
- composting
- constructing proper drainage
- planting trees as windbreaks
- cultivating crops in strips
- rotating crops
- planting cover crops (for example, sweet potato)

Activity 2.4.3 Information and communication technologies (ICT) and me: Erosion in my community

ATTENTION

Adult supervision required for this activity.

1. Work in pairs to determine an area in your community or near your school where erosion has occurred or is likely to occur.

2. Take a picture of the area. Write THREE paragraphs explaining why the erosion might have occurred at that place or could occur there. State some ways the issue could be addressed. Include the following in your explanation:

    - a picture of the site
    - a description of the area
    - TWO main causes of erosion
    - TWO possible solutions to the erosion issue.

Let's examine the causes and effects of pollution.

1. What are THREE major types of pollution?

2. What kinds of pollution do you see around your community? Give TWO examples.

2.5 Overfishing

The Pedro Bank, located off the south coast of Jamaica, is the largest and most valuable fishing ground for Jamaica's fishermen. Unfortunately, this is an area where overfishing takes place.

Overfishing occurs when too many fish are taken from an aquatic environment. Taking more fish than can be replaced naturally harms these organisms and their environment.

Overfishing removes the young fish stock that the fishing industry needs. Young fish grow into adults, which then produce more fish. Removing young fish reduces the adult fish population. Fewer fish means higher fish prices.

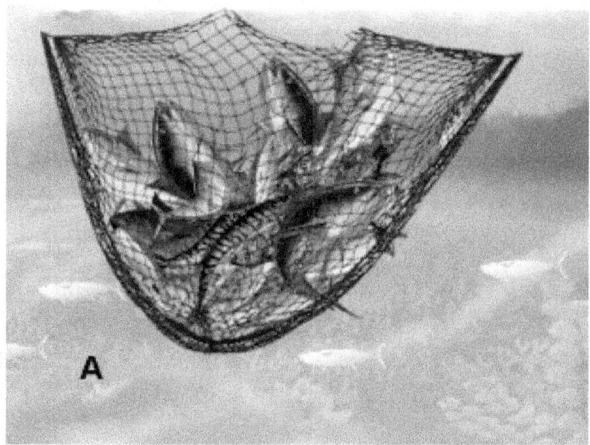 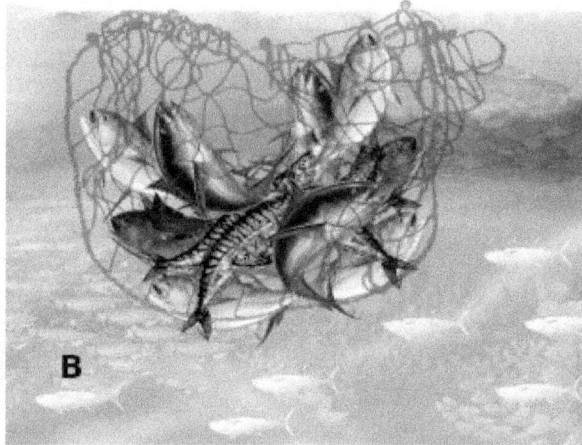

Figure 2.22: Overfishing.
A. Fishermen haul a net with small holes that capture baby fish.
B. Fishermen haul a net with big fish while smaller fish escape.

Laws help to prevent overfishing. One of these laws is regulating the sizes of fishing nets, so baby fish are left in the sea. The government also announces special fishing seasons for conchs and other shellfish. This helps with overfishing, as fishers can only catch them during those times of the year. These measures make sure aquatic food stocks survive.

In Jamaica, the National Environment & Planning Agency (NEPA) has strict rules about the fishing industry. NEPA also manages the seaside fishing resources. Learn more at:

https://websitearchive2020.nepa.gov.jm/new/legal_matters/laws/Environmental_Laws/Fishing_Industry_Regulations_1976.pdf

Did you know?

These animals and plants in Jamaica are almost extinct:

yellow-billed parrot

coney

iguana

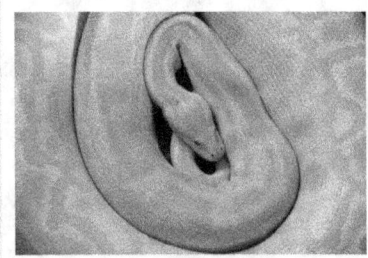
yellow snake

doctor bird

orchids

ferns

crocodile

2.6 What can humans do to protect the environment?

Environmental conservation means protecting the planet and preserving its natural resources. There are lots of things humans can do to take care of the Earth. Some of these environmental conservation practices are shown below.

Environmental conservation practices

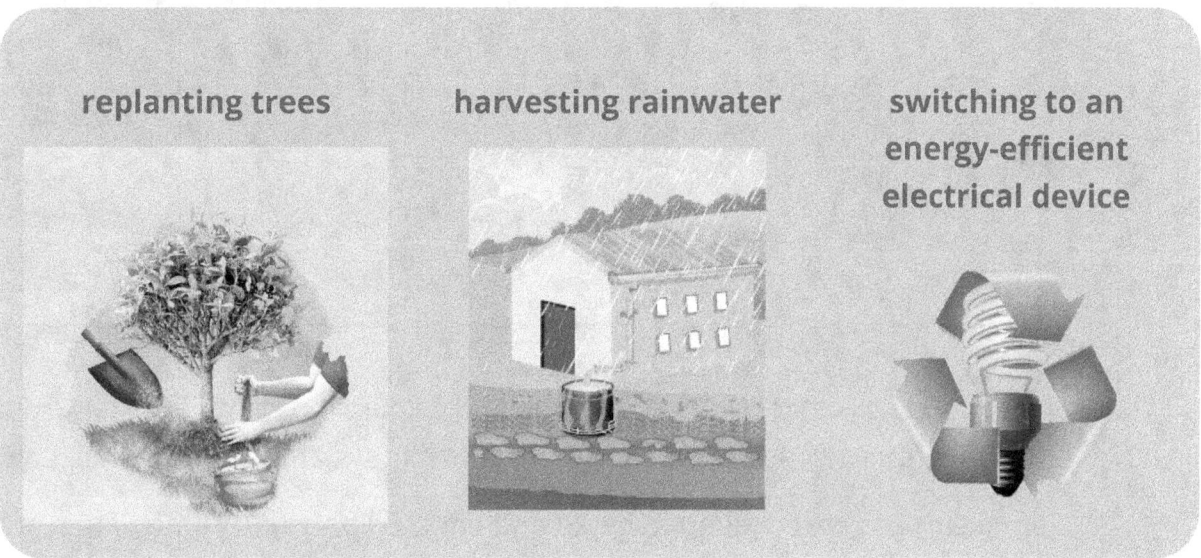

replanting trees

harvesting rainwater

switching to an energy-efficient electrical device

composting

image of kids thinking about how to conserve the environment

sharing their knowledge of climate change with others

conserving natural resources

(underground water (rivers and seas), land (soil), watershed and forests)

using natural and environmentally friendly products.

(will not harm the environment)

creating policies to prevent overfishing and the dumping of waste in the sea

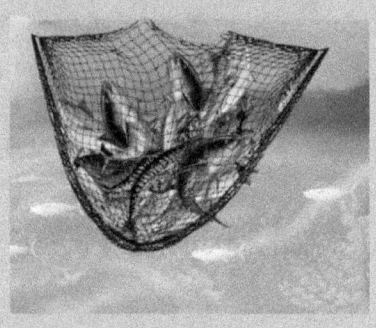

Did you know that smuggling plants and wildlife into Jamaica can introduce dangerous diseases and create situations that damage our own species?

Jamaicans are encouraged to catch and eat the lionfish to remove this invasive species significantly decreasing local fish populations in our rivers.

Jamaica taking action against invasive alien species – Jamaica Information Service (jis.gov.jm)

Activity 2.6.1: Information and communication technologies (ICT) and me: The effects of humans on the environment

Conduct research on the internet to find pictures showing some ways in which humans have harmed the environment. The pictures can include activities such as littering, burning garbage or polluting the land, water or air.

Use an app, such as *Animoto* and add music or sounds to the pictures to create a video entitled, 'The Effects of Human Beings on the Environment'.

Activity 2.6.2 Conservation

Using these four bins, consider this slogan: *'Reduce, reuse, recycle'*. Create a poster to support the slogan.

Activity 2.6.3 Science, technology, engineering, arts and mathematics (STEAM) and me: A problem in my environment

- ✓ Examine the community around your home or school. Find a problem that is affecting the environment you live in.
- ✓ Take a picture of the problem.
- ✓ Create a model to show the problem. For example, you can use cardboard, clay, pebbles, plastic and water to create a model of river pollution.
- ✓ Create and write a plan to solve the problem.

CHEETAH™ collaboration corner

Activity 2.6.4 Conservation connection

Part 1

Work in groups, as directed by your teacher, to think of some ways that you can conserve the environment. Include them in a short story or poem to inspire others.

Part 2

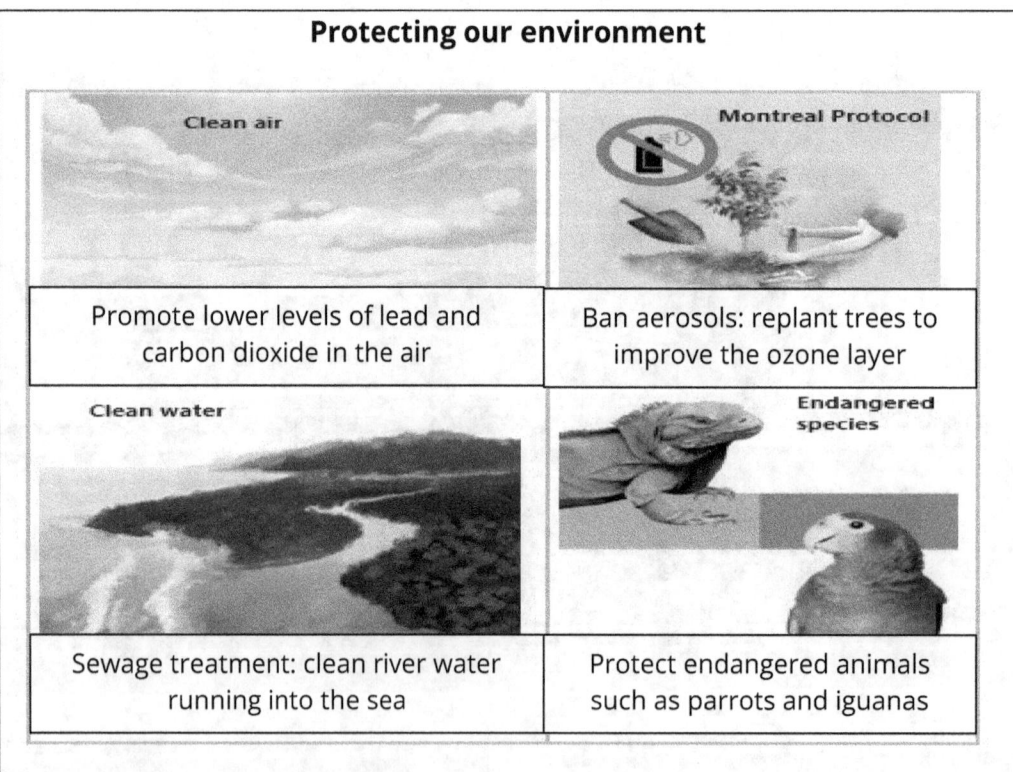

Look at the images on the previous page showing how the environment can be protected. Research to find out more about these methods of protecting the environment, then answer the following questions:

1. Why should we protect our environment?

2. List FOUR ways to protect the environment.

3. What is meant by endangered species? List FOUR endangered species in Jamaica.

4. What TWO things would you do to ensure that sewage does not contaminate our drinking water supply?

5. Give TWO laws we have in our country that protect the environment.

Grab your Merge Cube to explore the augmented reality activities next.

2.7 Augmented reality

Using the Merge Cube: The Merge Cube contains stored information that activates augmented reality apps. To use it, you just hold your Merge Cube in front of the augmented reality app for each activity.

Activity: Environmental adaptation

Organisms have to find ways to survive in augmented reality in their environment. Often, the ones most suited to the current environment will thrive. If that environment changes, those organisms might have to change as well or they could go extinct.

Activity: The human effect

In the past 200 years, the Earth's population has exploded and humans have mastered increasingly complex machinery. Humans have learned to build factories and operate large vehicles on dry land and water. By doing these things, we have changed the landscape all around us. Human activities have polluted the oceans and the air and destroyed entire habitats through deforestation. In these modules, you will see the impact of human activities first-hand and learn what all of us can do to make positive changes. Learn more on your Merge Cube.

Evaluate yourself!

Use this evaluation grid to check your understanding of the concepts discussed in this chapter. Read each statement below and insert the symbol that best shows how well you feel you understand the concept. Ask a teacher or parent to help you go over any areas that are still unclear, or that you do not feel you have mastered. Be honest!

I got it!

I need to do more work.

I do not get it. I need help.

In this chapter:

	I got it!	I need to do more work.	I do not get it. I need help.
1. I can outline the effects of human activities on the environment.			
2. I can describe the causes of soil degradation and explain how to prevent it.			
3. I can explain how soil degradation can be prevented.			
4. I am concerned about the impact of humans on the environment.			
5. I am aware of our responsibility to preserve the environment.			

Let's go, let's go! Let's hop over to the next chapter! We need to leap for more knowledge!

Chapter 3: Soil and climate change

Chapter objectives

- Investigate features/soils of different environments.
- Formulate a simple working definition of climate change.
- Use evidence from everyday local/regional/international situations to explain the effects of climate change on humans.
- Explore ways of reducing factors that cause climate change.
- Show concern for the impact of environmental problems on humans.
- Be aware of their responsibility to carry out good environmental practices.

CHEETAH Science Fiction

Mr Hiro's Machine

'I wish Mr Hiro could have been here today,' I think to myself, as I creep toward the massive machine in Mr Hiro's garage. Seven fist-sized drones buzz around me like bumblebees, their tiny cameras fixed on the cold, black cube in my arms.

The entire world waits with bated breath, hoping for a solution. Outside, electric thunder booms, and something gigantic crumbles and crashes into the ocean as the climate around our small island worsens. The garage lights flicker. Global warming is taking its toll.

A day before his death, Mr Hiro, my brilliant science teacher, had called me into his office. 'Dwayne, are you up to the task?' he had asked after explaining his plan, giving me a small cube to start the machine and instructing me on how to start the machine to stop global warming. 'Yes, I am,' I had replied.

Mr Hiro built this computer, the smartest machine in existence, to reverse the damage done by man to the environment and stop climate change once and for all. The machine has the power to remove carbon dioxide from the air and convert it to sugar. It is called a Photosynthesiser.

I spot the square hole in the computer's large glass casing. In a few seconds, the new age will begin. No melting glaciers, flooding, droughts, diseases and famine caused by global warming.

My hands tingle and my arms shake, but I pretend I'm an unfeeling robot, and ignore the sensations. I move the starter cube towards its dock. I insert the cube. It clicks and begins to hum. I stand back. The machine continues to hum, the suction valves open, the suction fans begin to spin and neon green lights flicker. My eyes dart between the uncountable parts behind the glass wall as they twist, turn and pulsate. I watch in fascination and await the outcome.

Something else happens. The cube opens like a flower and pushes a strange object upwards on a small platform. I can't figure out what it is. It's shaped like an oversized walnut. The object stops moving and the whole machine lights up with the colours of the rainbow. I can hear the gush of water swirling around inside the machine. The machine speaks.

'Hiro to the rescue.' It sounds like a thousand human voices talking in sync. 'What can I do for you today?'

One of the drones zooms toward my mouth. I clear my throat and I speak into the floating microphone. 'Hiro, save us from global warming and climate change.' I press the button, as Mr Hiro had instructed before his death, and wait.

3.1 Soil

Figure 3.1: *Plant rooted in the soil*

Soil is an important part of the environment. All food chains start with plants as the producers. The soil provides the right habitat for plants to grow. Animals (herbivores) eat plants for food, while other animals (carnivores) eat herbivores for food. Humans, like animals, also eat herbivores for food. Without soil, a food chain like the one that follows would break down completely. If there is no soil, then there will be no food for animals to eat.

Six major types of soil

Soil is broken down into types based on texture and what it is made up of. Each type depends on the conditions at its location, such as the types of rocks and plants found there, the length of time that the soil has been forming, and the climate. Here are some of the main types of soil:

Table 3.1: Types of soil

Type	Description	Image
Sand	Sand is light and contains very little matter that has come from living things. Sand does not hold many nutrients. It contains fine stones with large air spaces. Water drains from sandy soil easily.	
Peat	Peat is a dark-coloured soil. It is found at the bottom of bogs, swamps and marshes. It generally contains a large amount of decomposed material from living things. Peat holds water for a long time.	

Type	Description	Image
Clay	Clay is heavy soil and is mainly found on riverbanks. It feels sticky when wet. It is very rich in minerals. Clay soil does not allow water to pass through easily.	
Chalk	Chalky soil (marl) is found in limestone and rocky areas. It is thin and stony, with a powdery texture. It has a small amount of humus. Water drains quickly from chalky soil and the organic material it contains decays quickly.	
Silt	Silt is made up of sand and clay (rock and minerals). The particles in silt are smaller than sand but larger than clay. This soil holds water easily. When it is wet, it is smooth and slippery like soap.	
Loam	Loam is mostly made of sand and silt, with some clay, along with rotting materials such as leaves and insects. It has fewer nutrients than clay and holds water better than sand.	

All soils do not fit neatly into the sand, silt or clay category. When soil is made up of a mixture of different types, we call it **composite soil**. Loam is an example of a composite soil type. It is made up of different mixtures of sand, silt and clay, which results in *sandy-loam* and *clay-loam*.

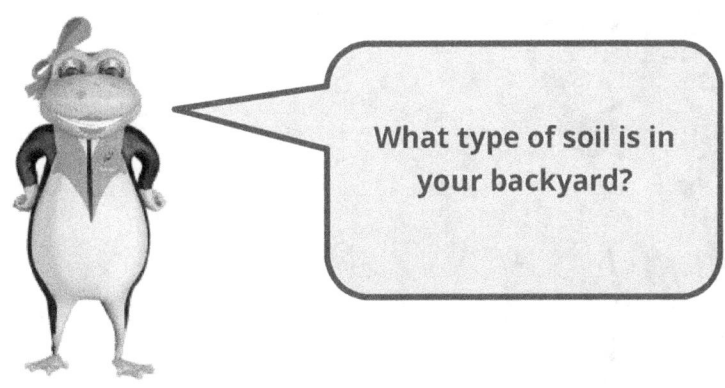

What type of soil is in your backyard?

Figure 3.2: Tractor preparing land for planting (tilling).

Figure 3.3: Farmer sowing seeds for crops.

One way that scientists determine what soil is composed of is by using a drainage test to see how quickly water drains through. This test is called a **percolation test**. Healthy soil will hold some water, but also drains any extra water easily. Living things also need air to survive. If the soil is filled with water, there is not enough space for air to pass through, and so plants will not get the air they need to survive.

Did you know? According to the Agricultural Land Management Division, there are 176 different soil types in Jamaica.

CHEETAH® PREPARING THE JAMAICAN SCIENTIST

 **CHEETAH™ collaboration corner**

Activity 3.1.1: Investigating soil

Part 1: Soil texture

Materials:

- three soil samples (sandy soil, clay soil and loam)
- magnifying glass
- table
- sheet of paper

Procedure:

1. Place a small amount of each soil sample on separate sheets of labelled paper.
2. Use your finger to feel the soil.
3. Use a magnifying glass to look at the soil crumb.
4. Record your observations and state any differences observed between the soil samples.

Do your observations agree with the following?

1. Sand feels gritty.
2. Loam feels smooth and is easily squeezed into a powder.
3. Clay feels sticky.

Part 2: Separating soil into components using the soil profile method

Materials:

- water
- clean, empty, transparent bottles with lids
- three soil samples from three different locations
- ruler

Procedure:

1. Label the three bottles as soil samples A2, B2 and C2.
2. Place equal amounts of each soil sample into its labelled bottle. Each soil sample should fill one-third of its bottle.

3. Pour water into each bottle to about three-quarters of its capacity. Leave some air at the top, as shown in the diagram.

4. Screw on the lid of each bottle tightly. Shake each bottle for two or three minutes. The water and soil in each bottle should be well mixed.

5. Allow the bottles to sit for one day or until the water is clear.

Observation:

1. How many layers do you observe?
2. Measure the thickness of each layer.
3. Create a table to record the thickness of each layer for each soil sample.

Here is an example of a possible observation:

Here are samples being measured from three different locations.

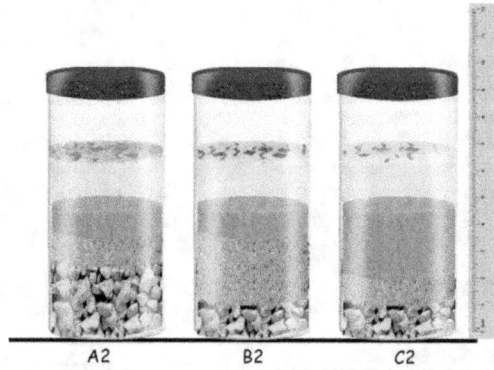

Do you see that each layer is a different thickness for each sample?
Do you see the humus floating on top of your own experiment?

Part 3: Water-holding capacity

A group of three students should work together.

Materials:

- three soil samples (sand, loam and clay sand)
- three measuring cylinders
- cotton wool
- three strips of graph paper or 1 ruler
- three similar transparent containers
- water
- three funnels of the same size
- stop clock

Procedure:

1. **Note: If you do not have a funnel, you may cut a used plastic bottle, as shown.**
   Cut off the top part of each plastic bottle and remove the lid to use as a funnel. Keep the bottom of each to use as a cup, as shown in the diagram.

 funnel    base of plastic bottle used as a cup

2. Label the funnels *sand*, *loam* and *clay*, then cork the funnel with cotton wool.
3. Put an equal amount of each sample into its labelled funnel.
4. Put each funnel into a measuring cylinder.
5. Pour 100 cm^3 of water at the same time.

Note: If the apparatus is a plastic bottle, use a ruler or paste a strip of graph paper as shown below, so that the graduations can be read and compared to give the volume (cm^3 or ml) of the filtrate as a height.

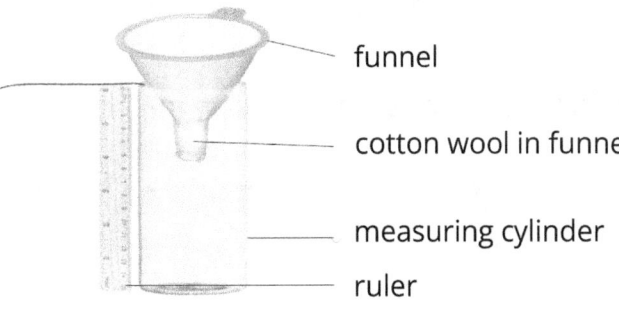

6. Put the same amount of soil into each funnel. Ensure that the soil is evenly distributed. Pour 100 cm³ of water over the soil in each funnel.

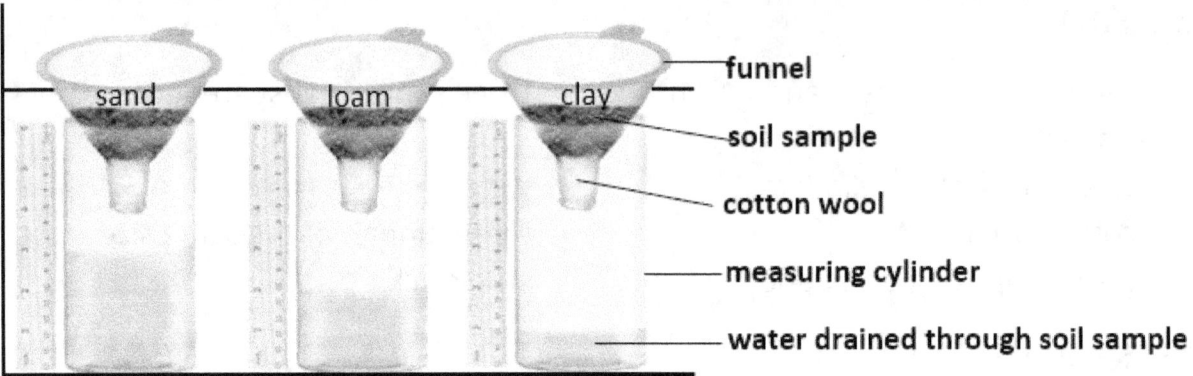

7. Record how much water is collected in each measuring cylinder at one-minute intervals over five minutes in the table below.

	After 1 min	After 2 mins	After 3 mins	After 4 mins	After 5 mins
Sand					
Loam					
Clay					

8. Compare the amount of water collected in the containers from each soil sample and discuss it within your group.

Now use your observations to answer the following questions.

1. Which soil held water for the shortest time?
2. Which soil held water for the longest time?
3. Which soil particles (larger or smaller) held the least water?
4. Which soil would be best for plant growth?
5. Which soil would be best for animals to live in?

Part 4: Air-holding capacity

Materials:

- soil samples of clay, sand and loam
- three small plastic cups labelled *clay*, *sand* and *loam*
- three large transparent containers of water labelled *clay*, *sand* and *loam*. These containers must be deeper than the cups.

Procedure:

1. Fill the small plastic cups with the soil samples as labelled. Knock the cup a few times on the table to compact the soil.

2. Immerse the plastic cups into the container of water as labelled.

3. Observe carefully to see which soil sample gives off the most bubbles.
4. Compare how many bubbles come from each soil sample.
5. Record your observation.

Now use your observations to answer the following questions.

1. Which soil has the most bubbles? What does that tell you about how much air is in the soil sample?
2. List your soil samples in order from the least air-holding capacity to most.

 Check your understanding of this section by answering these additional questions.

1. Why is soil important?
2. Your mother wants to start a vegetable garden in the backyard. Which soil do you think would work best for her? Why?
3. After investigating, you realise your mother does not have the ideal soil to grow vegetables, but you do not want to disappoint her. What actions would you recommend to ensure your mother's soil can grow her vegetables?
4. Explain how dead organic matter becomes manure for plants.
5. To improve soil quality, what would you suggest as a substitute for fertiliser?
6. Which sample has the most humus?
7. In which type of soil do you think you would most likely find plants with long roots growing deep into the soil? Explain why.

Did you know that there are more microorganisms in a handful of soil than there are people on Earth?

Microorganisms are tiny living things such as animals, dust mites and bacteria, algae and fungi that cannot be seen with the naked eye. You will need a microscope to view these.

3.2 Weather and climate

Countries and continents across the planet are affected by both weather and climate.

Weather tells us about daily changes in the elements of weather, which are temperature, air pressure, sunshine, rainfall, wind speed and wind direction at any place on Earth. These are the things the weather reporter talks about.

Climate is the expected weather seen in an area over a **long time,** say, about thirty years. There are three main types of climates found in different places on Earth: tropical or hot climate, temperate or cool climate and polar or cold climate.

Figure 3.4: Storm clouds seen from space.

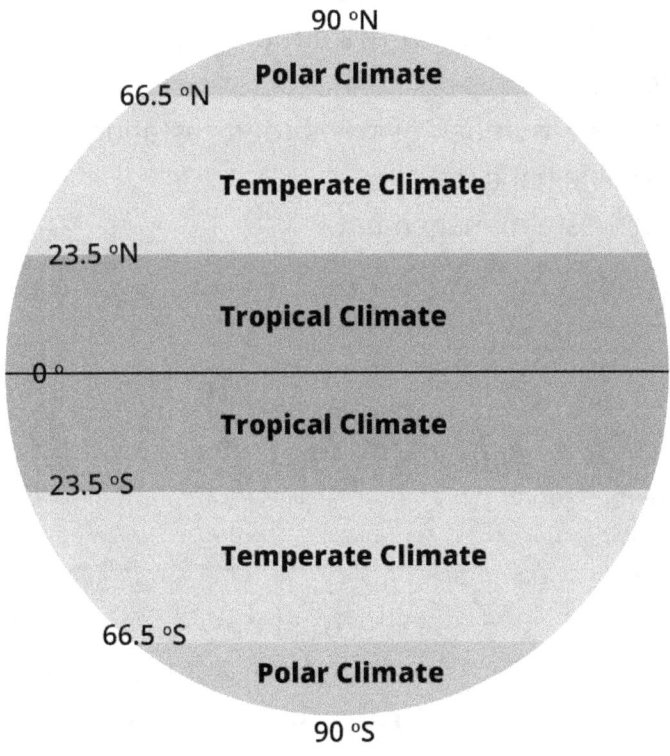

Figure 3.5: Earth's climates.

3.3 Climate change

Years ago, the weather pattern in each climate began to change and so the expected weather was not seen. Each climate was slowly measuring a higher temperature, becoming hotter. The watering holes were slowly drying up, and storms were becoming more violent.

Humans became worried and began to study climate science. They asked themselves what they were doing that might be causing these changes in the climate and what the effects were. This observation was called climate change.

Climate change is the change in Earth's overall average weather over a long time. Some of these changes are due to natural occurrences, such as volcanoes. However, most changes are caused by human activities.

Main causes of climate change:

1. burning fossil fuels gives off carbon dioxide
2. increasing temperature of the Earth's atmosphere (hotter temperatures)
3. burning plastics (e.g. PVC pipes)
4. cutting down a large number of trees (deforestation, which also leads to the destruction of wildlife habitats)
5. polluting by poor waste management.

Did you know all trees are plants but all plants are not trees? Some plants are shrubs (small plants that grow closer to the ground) while trees grow tall. Learn more by doing online research.

How will global warming affect sea turtles in the future?

The main effects of climate change due to the increased temperature are:

1. melting of ice caps on tops of mountains
2. flooding of low-lying areas
3. increasing sea levels
4. drying up of the Earth's water (droughts)
5. increasing food shortages
6. increasing tornadoes, earthquakes, hurricanes and wildfires.

Figure 3.6: *Melting of the ice at the poles of the Earth.*

Figure 3.7: *Climate change causes hurricane-force winds and a storm surge on the coast. The storm surge leads to flooding in low-lying areas.*

Figure 3.8: *A farmer uses drip irrigation to combat drought.*

Figure 3.9: *Forest wildfire.*

Did you know that a sea turtle's gender depends on how warm the sand is on their nesting beach? The warmer the temperature, the more female sea turtles are born.

How will global warming affect sea turtles in the future?

Reducing the effect of climate change

Burning fossil fuels (coal, petroleum and cooking gas) to get energy gives off carbon dioxide into the environment. To reduce carbon dioxide, we must turn to new forms of energy, such as wind energy, solar energy and hydroelectric energy.

There are many ways to reduce the amount of carbon that is produced on Earth:

1. conserving energy and water (turning lights and water off when not in use)
2. recycling waste
3. cycling, walking and carpooling
4. planting trees.

Did you know that Jamaica has a national tree-planting initiative that your school can participate in?

In 2015, over 200 countries agreed to cut their carbon dioxide emissions by looking after their forests, burning fewer fossil fuels and using more renewable energy.

Jamaica, like other countries, is taking steps to lessen the impact of climate change. There are new laws and policies to reduce pollution and improper garbage disposal. These new laws about managing water resources include collecting and storing water. There are also laws about adapting farming practices to withstand the changing environment.

Figure 3.10: *Solar panels.*

Jamaica has also started using renewable energy sources such as windmills, through the Wigton Windfarm Limited, and solar energy used in private homes, businesses and government institutions (**Community Access Points Being Retrofitted With Solar Panels–Jamaica Information Service (jis.gov.jm)**). The Jamaica Public Service, the major power-generating company in Jamaica, has eight hydroelectric plants. These convert the energy in flowing water into electrical energy. One of them is the Maggoty Hydroelectric Power Plant (Figure 3.11) in St. Elizabeth.

Figure 3.11: *The Maggoty Hydroelectric Power Plant.*

Using wind energy also reduces the dependency on fossil fuels.

Figure 3.12: Munro Wind Farm, St. Elizabeth, Jamaica.

The strategies to use more renewable energy sources and rely less on fossil fuels will decrease Jamaica's contribution to climate change. However, there is much more to be done.

Let's look at the causes and effects of climate change.

1. What are THREE causes of climate change?
2. What are THREE effects of climate change on the environment?
3. Give an example of how climate change can affect farmers.
4. How would you design your home to counter the effects of climate change?
5. How do you think climate change will change the January temperature in Jamaica?

Activity 3.3.1: Class discussion - Climate change in the Caribbean

Work with a partner to research and answer the following questions. Share your responses with your classmates.

1. Give an example of a Caribbean island greatly affected by climate change. Why is this island so vulnerable to the effects of climate change?

2. What can Caribbean islands do to reduce the impact of climate change?

3. Is there an agreement among the Caribbean islands about climate change? If so, why is it important to create this agreement? If not, why might it be difficult to create a climate agreement among the Caribbean islands?

4. View this website to see what measures the Jamaican government is taking, https://jis.gov.jm/features/jamaica-playing-its-part-to-combat-climate-change. What do you think of these measures?

Activity 3.3.2: Information and communication technologies (ICT) and me: Jamaica's changing climate

Use the internet to research the climatic changes taking place in Jamaica. Find the parish with the most average rainfall over the thirty years (1971-2000). Create a bar graph showing the average rainfall in August for the parishes with the two highest average rainfall and the two lowest average rainfall for the month. What conclusions can you make about the climate conditions in those parishes?

Be sure to use reliable sources, such as http://metservice.gov.jm/

Let's examine ways to prevent climate change.

1. What are TWO major ways to reduce climate change?
2. What are THREE things we can all do to help reduce climate change?

CHEETAH™ collaboration corner

Activity 3.3.3: Public service announcement

What can you do TODAY to reduce climate change?

You will work in pairs to create an online public service announcement. You want people to know about how humans impact the climate. Create TWO digital posters that can be printed and displayed or emailed to others, such as the examples shown.

Discuss the following guidelines together and plan your campaign before you begin:

- Have a theme that runs across the two posters to show they are part of the same campaign.
- Show the causes and effects of two separate issues.
- Focus on how humans may be harming the environment.
- Call on people to make a change.

Each person should then take responsibility for creating one of the two posters. As you work, regularly come together to check that the posters are clearly part of the same campaign. Each poster should share the messages that you had agreed on. When you have completed the posters, print and hang them in a visible area of your school or community, as well as email them to friends and family.

Human impact and climate change

- chemical erosion
- climate change
- crop rotation
- deforestation
- exhaustion
- organic agriculture
- overfishing
- overgrazing
- pesticide
- pollution
- reforestation
- slash-and-burn
- soil conservation
- soil degradation
- soil erosion
- solid waste
- terracing
- vector
- water-borne disease

```
Z P I S O I L C O N S E R V A T I O N I U T Q H Z
C L G U L W R E F O R E S T A T I O N Y O V Z B L
R V B M E C N J J S O I L E R O S I O N Q Y T G D
O J D U Q J K H Z K I V X W F Q K U V E C T O R M
P G E V X O R G A N I C A G R I C U L T U R E J P
R C G B D K A B E D S V J E M H W P I R A U Q R O
O K A L J R Q O S R G S P U I G M Y M H A I P S L
T B J N A Z J H F Z T S Z E D I T R S U U Z S O L
A W J N C L I M A T E C H A N G E P R O C R N V U
T M U Y U R B A N I Z A T I O N T H V X A W Y R T
I X U Y H L H S M M L V E B A D B E U E F X A G I
O A O V E R F I S H I N G A D K T H W S Q C N L O
N M H A S O V E R G R A Z I N G B W F U T O S Y N
S B P I X J M C T J S O I L D E G R A D A T I O N
R U U S T W X I V Z Z E T E R R A C I N G G Z W T
```

Evaluate yourself!

Use this evaluation grid to check your understanding of the concepts discussed in this chapter. Read each statement below and insert the symbol that best shows how well you feel you understand the concept. Ask a teacher or parent to help you go over any areas that are still unclear, or that you do not feel you have mastered. Be honest!

I got it!   I need to do more work.   I do not get it. I need help.

In this chapter:

		I got it!	I need to do more work.	I do not get it. I need help.
1	I can define climate change.			
2	I can explain how climate change and pollution affect our lives daily.			
3	I can use evidence from everyday local /regional/international situations to explain the effects of climate change on humans			
4	I can look at ways of reducing factors that cause climate change and solid waste pollution.			
5	I am concerned about the impact of environmental problems on humans.			
6	I am aware of my responsibility to carry out good environmental practices.			

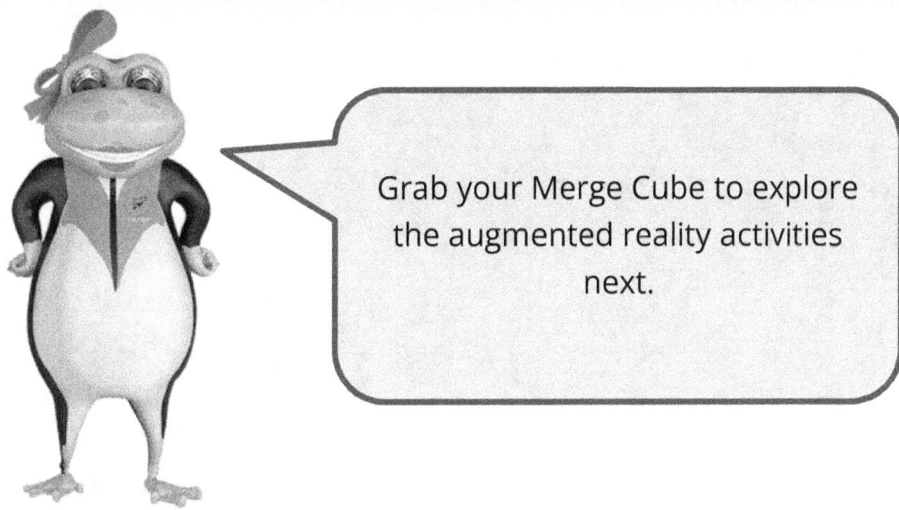

Augmented reality: Water and weather

In this experience, we'll look at the atmosphere and its different zones. The **atmosphere** extends far beyond the surface of the Earth and some scientists even say it goes all the way to the moon.

Activity: Layers of the atmosphere

Earth's **atmosphere** is a layer of gases that surround the planet, all held in place by Earth's gravity. We call it air, but its density (thickness) changes as you get farther from Earth. Use this module to visualise the different altitudes of the atmosphere and how we label them.

Activity: The water cycle

Water continually cycles around the land, the ocean and in the atmosphere via evaporation, condensation and precipitation. View this interactive water cycle to visualise these effects!

Activity: Extreme weather

In this module, you can view different forms of weather on the Merge Cube. View examples of blizzards, thunderstorms, droughts and temperature. Other examples are extreme tornados, landslides and hurricanes.

Term 1, Unit 2: Light

Words worth knowing:

The list below contains the scientific vocabulary that will be discussed in this unit.

- absorption
- artificial light
- concave lens
- convex lens
- curved mirror
- incident rays
- intensity
- lens
- light
- loudness
- luminous object
- magnifying glass
- microscope
- mirage
- natural light
- noise pollution
- non-luminous object
- opaque object
- periscope
- plane mirror
- rainbow
- ray
- reflected rays
- reflection
- refraction
- regular reflection
- shadow
- sound
- telescope
- translucent object
- transparent object

Remember to go back to the beginning of this unit and complete the table with the objectives by placing a tick in the box beside each objective that you have confidently achieved.

Chapter 4:
Light in our world

Chapter objectives

Specific learning objectives:
- ✓ Describe the properties of light.
- ✓ Investigate the properties of light.
- ✓ Distinguish between luminous and non-luminous objects/organisms.
- ✓ Investigate the interaction of light with materials that are shiny, dull, transparent, translucent and opaque.
- ✓ Investigate the interaction of light with lenses/mirrors.
- ✓ Investigate some effects of reflection/refraction in everyday life.
- ✓ Carry out fair tests in conducting investigations on the properties of light.
- ✓ Show objectivity by using data and information to validate observations and explanations about light.

CHEETAH Science Fiction

The Fastest Runner

Hooty Hoot was sitting in a tree, staring up at the night sky. LaChase was resting on a branch beside her, boasting about his impressive running speed.

'I'm so quick,' said LaChase. 'Nothing can move as fast as me!'

Hooty Hoot laughed. 'There are many things that can move faster than you!' she said.

LaChase was so shocked; he almost fell off the branch. 'Name one!' he demanded.

'I'll do better than that,' said Hooty Hoot. 'I'll show you the fastest thing in the entire universe.' And with that, she pointed to the glowing moon.

'The moon isn't as fast as me,' laughed LaChase.

'Not the moon, silly. I'm talking about the light that reflects off it.' LaChase scratched his head.

'The moon doesn't make its own light, LaChase. The sun illuminates it, and that's why it appears so bright in the night sky. Light travels at 300 million metres per second through space. And nothing moves faster than that.'

'Nonsense,' said LaChase. 'Light isn't faster than me, and I'll prove it by beating it in a race!'

'Challenge accepted,' said Hooty Hoot, rummaging in her nest for a flashlight. The starting line was the tree where Hooty has her nest.

'First one to the rock wins the race,' said Hooty Hoot, pointing to a huge rock across the clearing. 'I'll even give you a head start.' But LaChase was already running as fast as he could across the grass, heading for the finishing line at the rock.

Hooty Hoot aimed the flashlight at the rock. When LaChase was halfway to the finish line, she flicked the 'on' switch. Light flashed from the bulb and lit up the rock in an instant. When he saw the light on the rock, LaChase collapsed on the grass, puffing and panting.

'Ok,' said LaChase. 'Light does travel faster than me.'

4.1: What is light?

Figure 4.1: Surface of the sun showing solar energy.

Light is a form of energy that is visible and travels in the form of **rays**. Light *rays* are waves of light that travel in straight paths. A group of **rays** that move in the same direction from a light source forms a beam. Our eyes need light to see - we cannot see anything if it's completely dark.

On Earth, much of the natural light we use comes from the sun in the form of heat and light. The light energy emitted (given off) from the sun travels at 300 million metres per second across outer space. Nothing in the universe moves faster than light.

Did you know that the sun's light takes about 8 minutes and 20 seconds to reach us on the Earth?

Properties of light

Some characteristics of light are that it:

- travels in a straight line
- bounces off objects (reflection)
- bends when it passes from one type of material to another (refraction).

Activity 4.1.1: Investigating a property of light

In this activity, you will examine one property of light.

- one straw (straight)
- one lit candle

Procedure:

1. Use a straight straw to look at a candle flame (keep the straw at least 20 cm from the flame).
2. Bend the straw slightly but do not break it.
3. Use the same bent straw to look at the same flame again from the same distance as before.
4. Observe and record your observations.

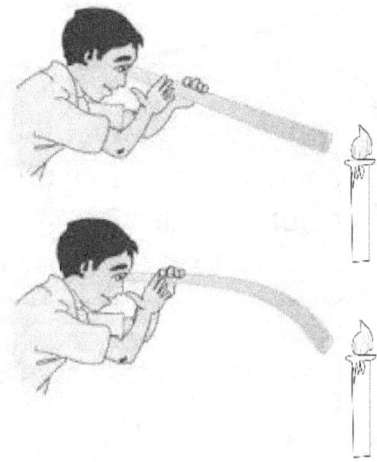

Questions:

1. Explain why the flame could or could not be seen using the bent straw.
2. Give reasons for your observations.
3. What conclusion can you draw from your observation?
4. State the property of light investigated in this activity.

Activity 4.1.2: Investigating a property of light

In this activity, you will examine the property of light.

Materials:

- three pieces of cardboard, 10 cm by 10 cm each
- light source (flashlight)
- masking tape
- straw
- play dough or mixed flour
- table top

Procedure:

1. Pack the three pieces of cardboard together and punch a single hole in the centre of all three pieces.
2. Tape one end of the straw to the back of each piece of cardboard.
3. Align the holes and secure the cardboard using masking tape then put a flashlight on one end.
4. Push the other end of the straw into play dough or mixed flour so that the cardboards are upright and secure to the table.
5. Arrange the cardboard pieces so that they are about 15 cm apart from each other.
6. View the light from the flashlight from the other end through the holes in the cardboard.
7. While looking through the hole, shift the cardboard in the middle and observe the flashlight again.

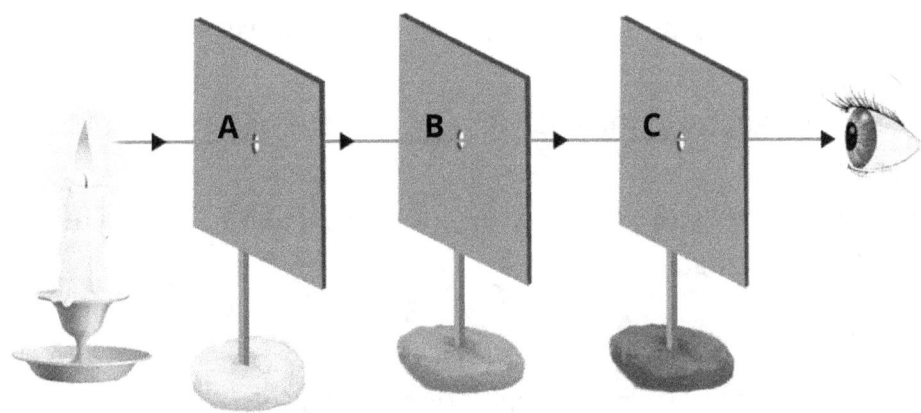

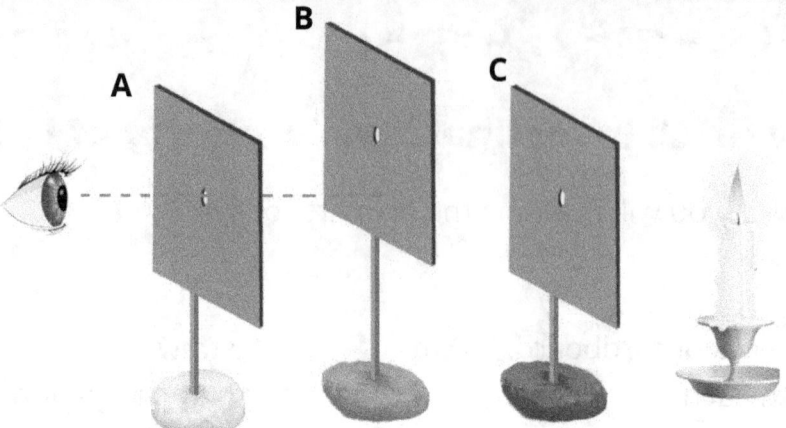

Record your observation.

What conclusion can you draw from your observation?

Activity 4.1.3: Investigating a property of light

In this activity, you will examine the property of light.

Materials:

- three pieces of cardboard 10 cm x10 cm each
- light source (flashlight)
- masking tape

Procedure:

1. Pack the three pieces of cardboard and punch identical holes in the centre of all three pieces.
2. Arrange the boards so that they are about 15 cm apart from each other.
3. Align the holes and secure the cardboard using masking tape and put a flashlight on one end.
4. View the light from the flashlight from the other end through the holes in the cardboard.
5. While looking through the hole, shift the cardboard in the middle and observe the flashlight again.

Record your observation.

What conclusion can you draw from your observation?

Natural light sources

All sources of light on Earth can be classified as either *natural light* or *artificial light*. The sun is a natural light source because it makes its own light. Other natural light sources include:

- the stars
- fire
- lightning
- animals that give off their own light (bioluminescent organisms) such as fireflies, glowworms, jellyfish and anglerfish.

Figure 4.2: Firefly.

Did you know fireflies produce light using a chemical reaction inside their abdomen?

Do you know the Jamaican nickname for fireflies? Ask your parents or teachers if you do not know.

Artificial light sources

Figure 4.3: A paraffin lamp - an artificial light source.

Artificial light sources are objects that are made by humans. These include:

- light bulb
- flashlight
- television
- firework
- fire
- lamp light
- candlelight.

Did you know that Thomas Edison is believed to have invented the light bulb and said, 'Now I know 10,000 ways not to make a light bulb.'

4.2: Luminous and non-luminous objects

Luminous objects are objects that give off or `emit' their own light. Luminous objects can be man-made or naturally occurring. Examples include the sun, stars, lit candles, lamps or matchsticks, fireflies, light bulbs and glowworms.

Objects that do not emit light are called *non-luminous objects*. Non-luminous objects reflect light from luminous objects. Examples of non-luminous objects include the moon, trees and books. In fact, anything that is not a light source is non-luminous.

Non-luminous objects can be transparent, translucent or opaque.

- *Transparent objects* allow light from natural and artificial light sources to pass through them. These objects are often made of clear or colourless materials. Some examples of transparent objects include plastic bottles, glass bottles or the lens of eyeglasses.

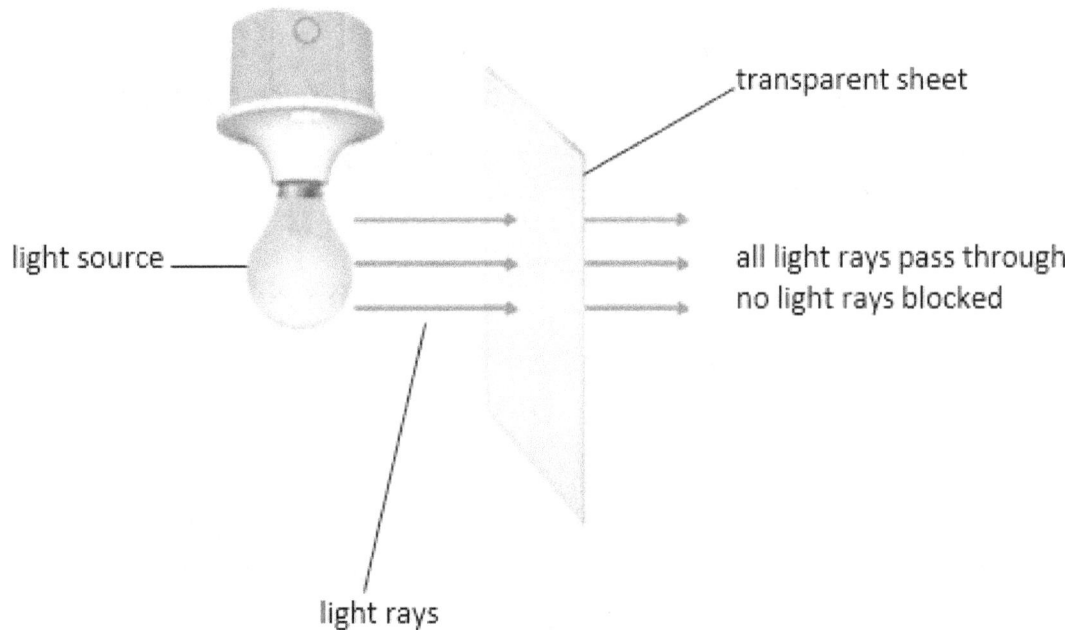

Figure 4.4 (a): *Light passing through transparent objects.*

- *Translucent objects* only allow some light to pass through them. Objects that are behind *translucent* materials appear blurred. Some examples of translucent objects are frosted glass, greased paper and some plastic bags.

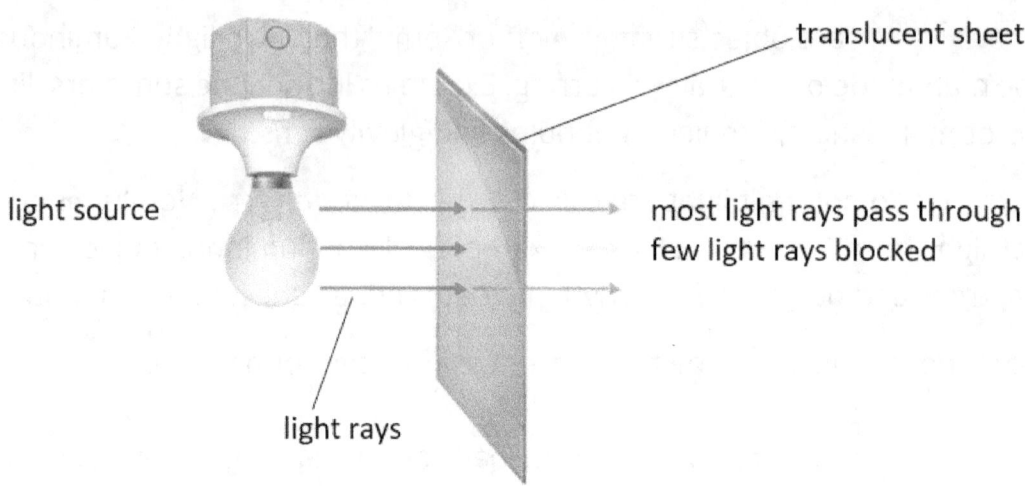

Figure 4.4 (b): *Light passing through translucent objects.*

- *Opaque objects* do not allow light to pass through them at all, so objects behind these materials cannot be seen. Opaque objects will therefore produce shadows whenever light shines on them. Examples of opaque objects include metal doors, walls, cardboard boxes and books.

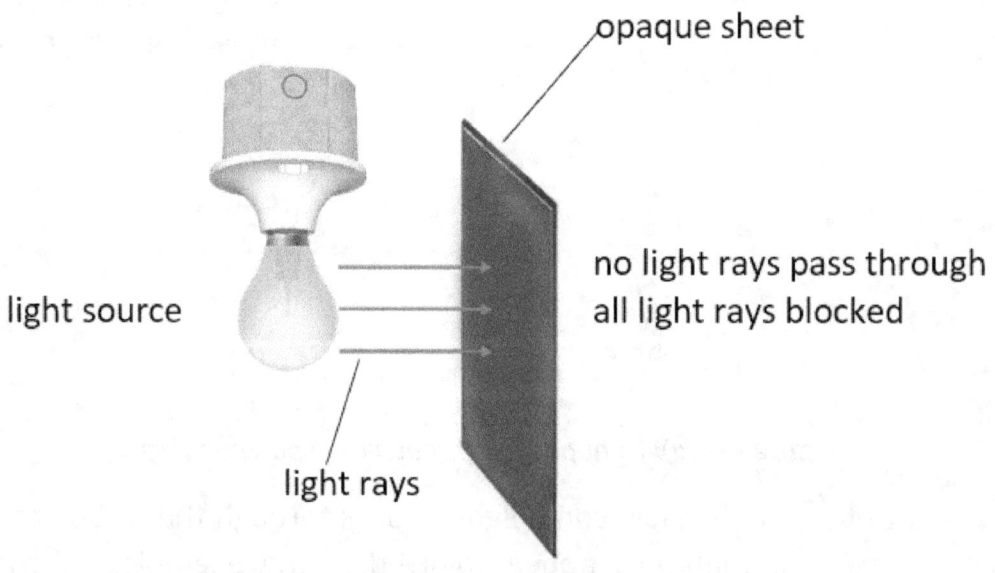

Figure 4.4 (c): *No light passes through opaque objects.*

Activity 4.2.1: Investigating the effects of light on objects

In this activity, you will examine the characteristics of some non-luminous objects.

ATTENTION
Adult supervision required for this activity.

Materials:

- six objects (e.g. large book, cardboard box, a mirror, black cloth and white cloth)
- plastic cups or plastic bags (clear, white and black)
- light source such as a candle or small flashlight

Non-luminous object (wooded door) positioned between an observer (man) and the light source (flashlight or candle).

Figure 4.4.1: *Light shone on a non-luminous object.*

Procedure:

1. Collect at least six non-luminous objects, such as a large book, cardboard box, a mirror, black cloth, white cloth, in addition to clear, white and black plastic bags.

2. Locate a light source such as a flashlight or candle to use in this experiment.

3. Set up an experiment with an observer behind the object, as shown in Figure 4.4.1, to investigate the effect of light on each non-luminous object.

4. Record your observations for each of your non-luminous objects in a table like this:

Objects	Transparent (all light passes through)	Translucent (some light passes through)	Opaque (no light passes through)

Conclusion: What did you learn from this experiment? Use the words transparent, translucent and opaque in your answer.

Let's examine the characteristics of non-luminous objects:

1. Give THREE examples of non-luminous objects.

2. Define the following terms and give an example of each: opaque, transparent and translucent.

3. What happens when no light passes through an object?

4. How is a translucent object identified?

5. Look at the direction of the arrows in the drawing showing three circles A, B and C. Tell which circle is the sun, which is the Earth and which is the moon.

Did you know that the moon is non-luminous? It shines because its surface reflects light from the sun.

4.3: Shadows

What is a shadow?

Shadows are made when the light is blocked by an opaque object. Light travels in a straight line and cannot go around an object. If an opaque object blocks the path of the light, a shadow will be formed.

Figure 4.5: *Shadows cast from the light of the setting sun.*

CHEETAH™ collaboration corner

Activity 4.3.1: Investigating shadow formation

Arrange yourselves in groups of four, select and cooperate with a group leader.

Material:

- six opaque objects of different shapes and sizes
- small light source
- wall or screen
- cartridge paper/cardboard

Procedure:

1. Cut a hole in a sheet of cardboard/cartridge paper.
2. Turn on the light.
3. Allow light to come through the hole of the cardboard or cartridge paper and shine on the object.
4. Make sure all the objects are held at the same distance from the light source.
5. Move the object closer and further away from the light source.
6. Observe and record your observations.

Questions:

1. Does the shadow have the same shape as the object?
2. Is the size of the shadow the same as the object?
3. Move the objects closer to the light. What do you observe?
4. How does the distance between the light and the object affect the size of the shadow?
5. How does the size and shape of the shadow change as you move the object closer and further from the light source?

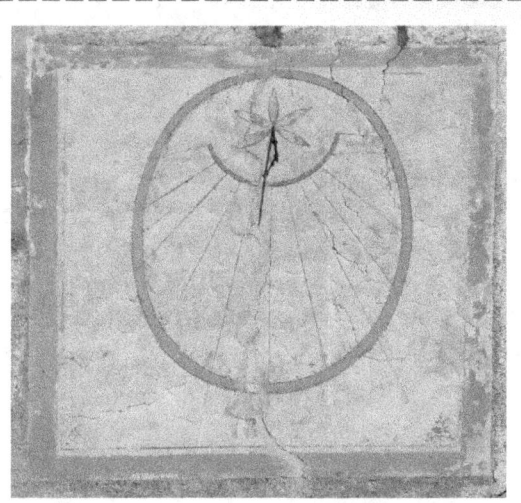

Ancient sundial

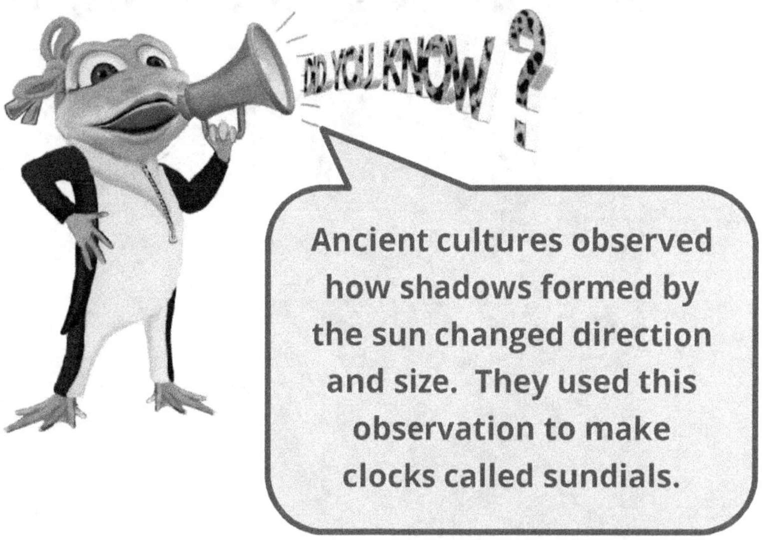

DID YOU KNOW?

Ancient cultures observed how shadows formed by the sun changed direction and size. They used this observation to make clocks called sundials.

Did you remember to teach someone at home what you know? Remember, we are preparing you to become a Jamaican scientist. As a scientist, you have a duty to share your knowledge with others.

4.4: Reflection of light

Figure 4.6: *Image of deer reflected in the water.*

What is reflection?

Earlier in this chapter, you learned that light does not pass through *opaque objects*. But where does the light go when it hits an opaque object?

When you look in a mirror, you see your reflection. *Reflection* occurs when a ray of light hits an object's surface and bounces off.

Figure 4.7: *Child's image reflected in the mirror.*

Types of reflection

Light is reflected by plane mirrors in two ways:

Regular reflection is the type of reflection that occurs when you look at a smooth, shiny, flat, and clean surface (known as a plane mirror). In regular reflection, light rays strike this surface and bounce off (reflected) in the same direction. This gives a clear image.

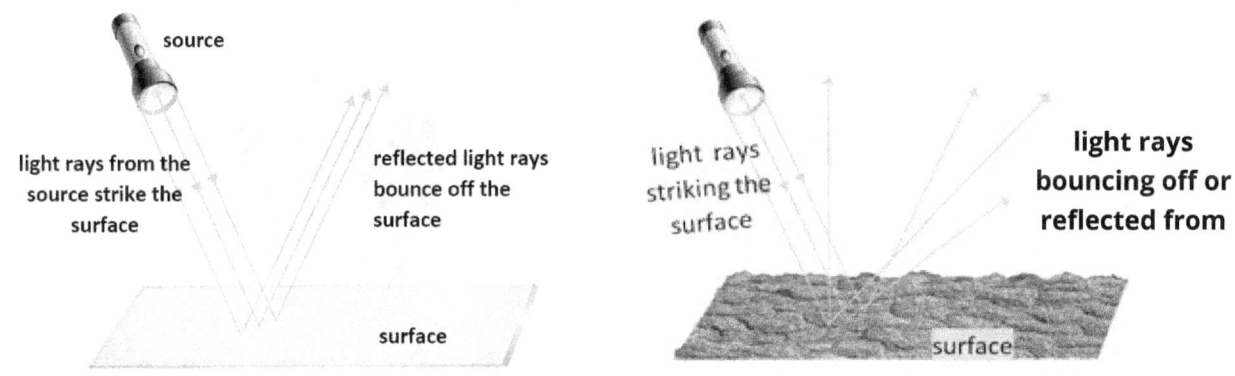

Figure 4.8: Regular reflection of light by a smooth, shiny reflecting surface (e.g. plane mirror).

Figure 4.9: Diffused light bouncing off a dull, bumpy reflective surface (e.g. wall).

Diffuse reflection is the type of reflection that occurs when light strikes a surface that is not flat and bounces off in different directions. For example, dusty car bodies and crumpled aluminum foil. Diffuse reflection forms blurred images.

Did you know?
Magicians use mirrors to give the illusion that objects have magically disappeared.

How are we able to see?

We can see this because of the following three steps:

1. Light from a source, like a light bulb, strikes an object, for example, a book.
2. The light bounces off the object.
3. The light from the object enters the eye.

Figure 4.10: Diagram showing the reflection of light rays from the book to the eyes. This reflection makes the book visible.

Without light, the object cannot reflect light. *Reflection* is the bouncing of light from an object when it strikes the object. Therefore, if there is no light, we cannot see an object.

Activity 4.4.1: Science, technology, engineering, arts and mathematics (STEAM) and me: Understanding how mirrors can be used to change the direction of light rays.

A **periscope** is a device that allows you to see over a wall or around a corner. It is mostly used in submarines so that crew members can see what is happening above the water without being seen. There are two types of periscopes: the see-behind and the see-in-front. It is the arrangement of the mirrors that makes the difference in the type of periscope. This is an example of how mirrors can be used in the direction of light.

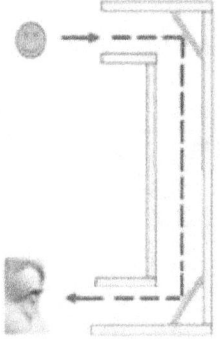

See behind periscope

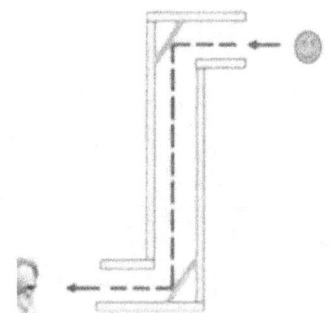

See in-front periscope

Questions:

1. Using the diagram above, explain how the periscope works. Include a description of how you used concepts from geometry to explain how the periscope works.

2. Describe how the arrangement of the mirrors affects the type of periscope present.

3. Write a short story or poem about how the periscope is used in the ocean.

4. This is a periscope. What do you think it is likely to be attached to? What do you think it is there to do?

Figure 4.11: *Periscope protruding from the water.*

Let's take another look at the reflection of light.

1. Describe what is meant by the reflection of light.
2. Give THREE examples of reflection in everyday life.
3. What type of surface is most likely to be able to reflect light?
4. State ONE difference between your face and the reflection you see in a mirror.
5. Why is light needed to see an object?

4.5: Refraction of light

What is refraction?

Refraction is when light travels from one material to another, causing the speed and the direction of the light rays to change. Different materials (media) bend light to different degrees. The thicker the material, the more it bends (refracts) the light.

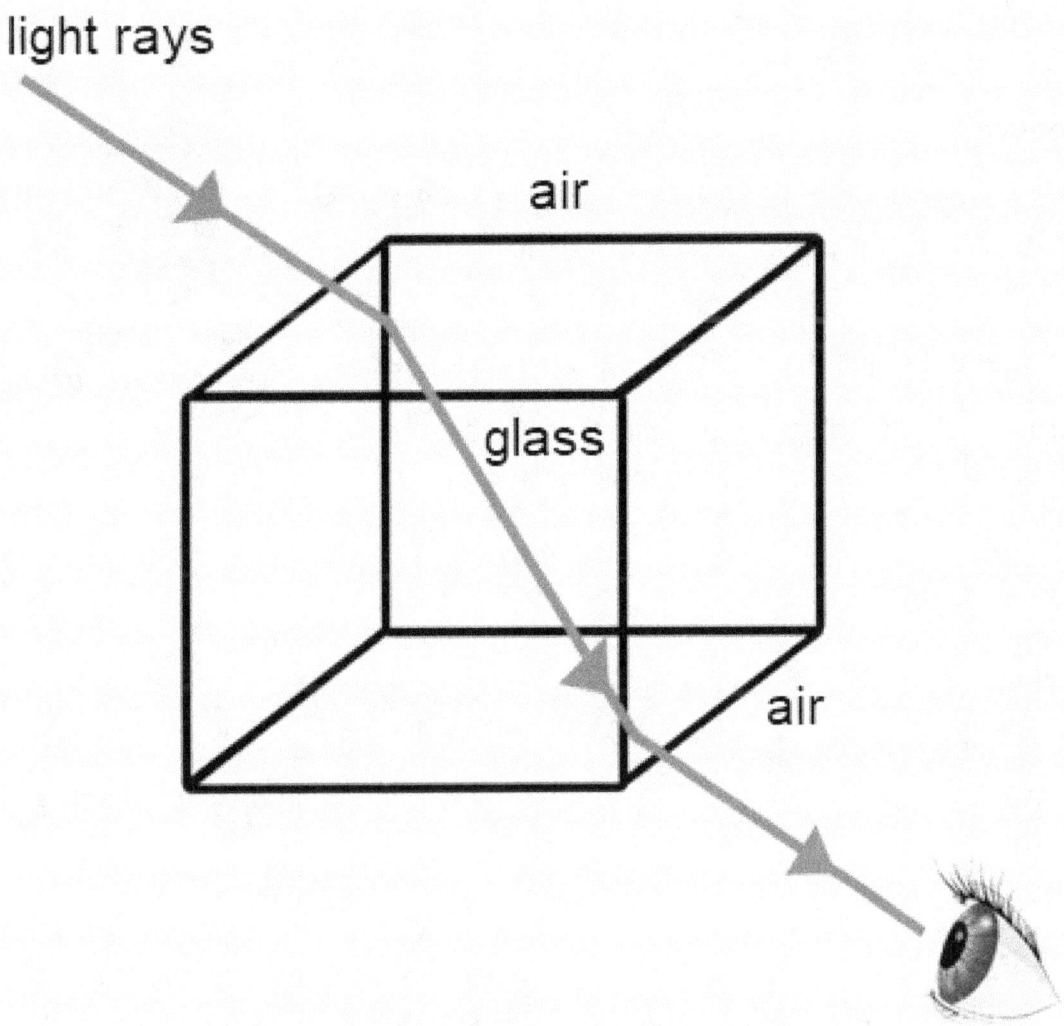

Figure 4.12: *Diagram showing how light rays bend as they pass from the air through a glass block.*

Activity 4.5.1: Investigating refraction

Work with a partner to investigate how the refraction of light affects how we see objects.

Part A

Materials:

- pencil or spoon
- 2/3 filled glass of water

Image showing refraction through the water.

Procedure:

1. Place a pencil or spoon (object) in a glass of water so that part of the pencil or spoon is shown above the waterline.
2. Look at the object from the side at various angles.
3. Draw a labelled diagram to record your observations. How did the object appear to look in the water?

Part B

Materials:

- opaque drinking glass
- water
- ruler
- coin

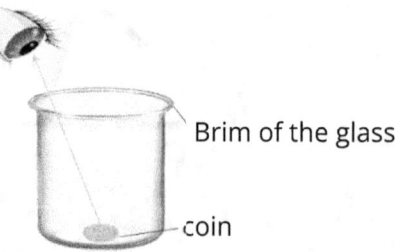

Diagram showing the original position of the eye looking at the coin in the centre of the glass bottom.

Procedure:

1. Place a coin in the middle of an empty opaque drinking glass that is sitting flat on a table. (If the glass is not opaque, you can wrap it with a piece of newspaper or construction paper.)

Diagram showing how refraction helps us to view the coin with the eye from a different position.

2. Start with your eyes looking directly downward at the coin in the glass. Move your head backwards and away from the drinking glass until the coin disappears from your viewpoint behind its rim, as shown in the diagram.

3. Hold your head in that position, while your partner slowly pours water into the glass.

4. Record your observations.

5. Switch roles and repeat the experiment.

Use the data you collected to answer the following questions:

1. What did you observe as the water was poured into the glass?

2. Why do you think the coin reappeared when you added water to the glass?

Refraction in our world

If you wear glasses, you depend on light refraction to see more clearly. Devices such as magnifying glasses, microscopes and telescopes use lenses to control how light refracts through them.

Figure 4.13: *Refraction through an eyeglass lens.*

What is a lens?

A *lens* is a transparent piece of glass or plastic with two curved surfaces. Lenses refract light and form an image. More curved lenses refract more light as the light passes through the lens surface.

Rainbows

A *rainbow* is an arch of seven colours (red, orange, yellow, green, blue, indigo and violet—R.O.Y.G.B.I.V.) in the sky created by the refraction of sunlight, as it passes through raindrops. For you to see a rainbow, the rain must be in front of you and the sunlight shining from behind you, so that refraction can take place. This causes the light to refract into your eye.

Figure 4.14: *Rainbow visible through the refraction of light.*

Mirages

Figure 4.15: *Mirage as a result of the refraction of light on layers of air.*

Mirages are images that look real but do not really exist. Mirages also occur due to the refraction of light rays. When layers of light pass through materials that are of different temperatures, they change direction. This can create a false image. Sometimes when driving along the road we see a pool of water in the distance but when we reach it, there is no water. This pool of water was a mirage. Sometimes desert travellers see an oasis in the distance but when they get nearer, there is no oasis. The mirage can create false images that seem very real.

CHEETAH® PREPARING THE JAMAICAN SCIENTIST

Did you know that 11%–35% of people sneeze when exposed to bright light in what is called a photic sneeze reflex?

CHEETAH™ collaboration corner

Activity 4.5.1: Information and communication technologies (ICT) and me: Surrounded by mirrors

Using the internet, research the use of flat and curved mirrors in everyday life. Create a portfolio of pictures of these mirrors in use in our environment, e.g. in bathrooms, on roadways, on motor vehicles or in stores. Explain how these mirrors help people to perform everyday tasks.

Activity 4.5.2: Light crossword puzzle

Across

1. surface that reflects a clear image

3. light rays bouncing off a surface

6. bend light to form an image

9. object that does not make its own light

Down

2. visible energy produced by the sun or light bulb

4. bending of light ray bends as it passes between mediums

5. object that makes its own light

7. reflective surface

8. non-reflective surface

10. where something comes from

Evaluate yourself!

Use this evaluation grid to check your understanding of the concepts discussed in this chapter. Read each statement below and insert the symbol that best shows how well you feel you understand the concept. Ask a teacher or parent to help you go over any areas that are still unclear, or that you do not feel you have mastered. Be honest!

I got it!

I need to do more work.

I do not get it. I need help.

In this chapter:

	I got it!	I need to do more work.	I do not get it. I need help.
1. I can describe the properties of light.			
2. I understand the investigations on the properties of light.			
3. I can differentiate between luminous and non-luminous objects.			
4. I can show the interaction of light with materials that are shiny, dull, transparent, translucent and opaque.			
5. I can investigate the interaction of light with lenses/mirrors.			
6. I can show examples of reflection and refraction in daily life.			

	I got it!	I need to do more work.	I do not get it. I need help.
7. I can do fair investigations on the properties of light.			
8. I can show objectivity by using data and information to prove my observations and explanations about light.			

Grab your Merge Cube to explore the augmented reality activities next.

Augmented reality: Instruments of light

Welcome to Instruments of light. In these activities, we will explore how light travels through different materials and why light can have different colours. Then we will learn how to manipulate light with a few nifty tools and perform some real neat tricks. Get ready!

Activity 1

Colours of light: Light is a wave, and its colour depends on the wavelength (the distance from one peak of the wave to the next). In this activity, become a light engineer and make your own waves.

Activity 2

Waves in space: Unlike sound waves which need to move through a substance, light can travel even in a vacuum. Try your hand at sending some sound waves and light waves shooting through space.

Augmented reality: Radiating light

In this card, we're going to look at light and how it illuminates the world around us. We'll also try to understand that light bounces around the world and its effects on how we visually perceive the world around us.

Activity 1

Illuminated world: Every object we see is really just light reflecting off of that object. In this module, you can see the inside of a dark cavern. Turn up the lights to see.

Activity 2

A world illuminating: Some objects emit their own light. Check out these mushrooms. By tapping on them, you can see what it's like when they emit their own light.

CHEETAH® PREPARING THE JAMAICAN SCIENTIST

Activity 3

Light passes through: In some cases, light can be blocked or dimmed by different materials. Close the shutters and lower the shades in the room to see how it affects what you can see.

Activity 4

Bouncing light: When we see objects of a certain colour, what we are really seeing is just that part of the spectrum of light reflecting off that object. In this module, click on different objects and see how the light only reflects the colour of the object to your eyes.

> According to Thomas Edison, 'There is no substitute for hard work.'
> Do you agree with him?

> Let's go, let's go, let's go! Let's leap for knowledge.

Chapter 5:
Sounds in our environment

Chapter objectives
Specific learning objectives:
- ✓ Describe sounds using appropriate scientific language.
- ✓ Investigate some properties of sound.
- ✓ Explain why sounds may be interpreted as pleasant/unpleasant.
- ✓ Identify sources of noise pollution and ways to eliminate them.
- ✓ Explain why loud sounds can be harmful.
- ✓ Formulate hypotheses when conducting investigations into the properties of sound.

CHEETAH Science Fiction

Sound

'Here comes Sound-Boy!' a student yelled as Jason walked towards the playground. Jason just rolled his eyes and kept walking to class. That one time he had shared his quirky superpower with his so-called friend was the last time he ever walked around without being teased. It had taken him a long time to understand why he could hear things that no one else could. How was it possible that he could hear the sand moving on the beach or hear things long before he saw them?

All the doctors and therapists his parents forced him to see, fearing he had a personality disorder, were baffled. They all tried to prove to him that what he said wasn't true. It was impossible. One doctor decided to play a song on his phone, very quietly, two floors below where Jason was waiting and asked him if he could hear it. The doctor's phone fell to the floor when Jason told him, 'You were listening to the happy birthday song. Weird choice, doc.'

Finally, after running special hearing tests, Jason learned that he could detect low sounds outside of the normal human range of hearing. He could hear sounds that only certain animals can hear.

Every day was an annoyance for Jason but a strange thing happened that day. As Jason sat in class, he heard a strange faint sound. It was a very low-pitched rumble that seemed far away. He didn't recognise the sound, but he knew it wasn't good.

'Miss, my head is going to explode! Don't you hear that noise?' he asked his teacher.

'What noise? I don't hear anything.'

'Yeah, I have very sensitive ears. I'm telling you, Miss, something is wrong.' Knowing about Jason's phenomenal hearing, the teacher told Jason to follow her as they fled to the principal's office. Minutes later, all the students heard the principal's warning over the intercom. 'Teachers, I want you to help the students to get away from the windows, bookshelves; anything that could fall or break. Go under the desks and tables and cover their heads with their hands. Please act now!'

As the building started to shake violently, Jason thought, 'I guess that's what an earthquake sounds like from a distance.' On that day, Jason became a hero and the teasing stopped.

5.1: Introduction to sound

What is sound?

Sound is anything that we can hear. A barking dog, music, clapping and talking are all sounds. Both sound and light are forms of energy. A sound is made when the particles of matter (anything that has mass and takes up space) move. Sound energy moves in all directions from the source (the vibrating object). Sound vibrations move energy from a vibrating object to its surroundings so we can hear.

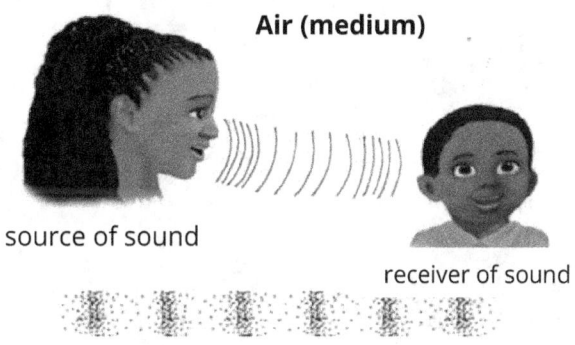

Figure 5.1: *Image showing sound waves moving through the air.*

As the sound moves, energy is lost as it strikes against objects. For instance, talking on a podium with a fan overhead will scatter the vibrations and people in the audience will not be able to hear. With the fan off, the sound energy will go further.

When you pluck a string on a guitar (Figure 5.2a), the string vibrates. The vibrating string pushes against the air particles next to it, which makes the air particles vibrate. These vibrations continue to spread through the air in all directions away from the guitar string and the energy in the vibration moves through the air. When vibrating air particles go into our ears, our brain interprets the vibrations as sound.

Figure 5.2a: *String instrument—guitar.* **Figure 5.2b:** *Wind instrument—trumpet.*

In Figure 5.2b, the air passing through the trumpet vibrates and creates a sound.

Sound needs a medium, a substance like a solid, liquid, or gas, to travel through. Vibrations move through these types of matter to carry sound. For example, the humpback whale can make sounds that travel thousands of miles through the water, which acts as the medium.

Figure 5.3: Humpback whale.

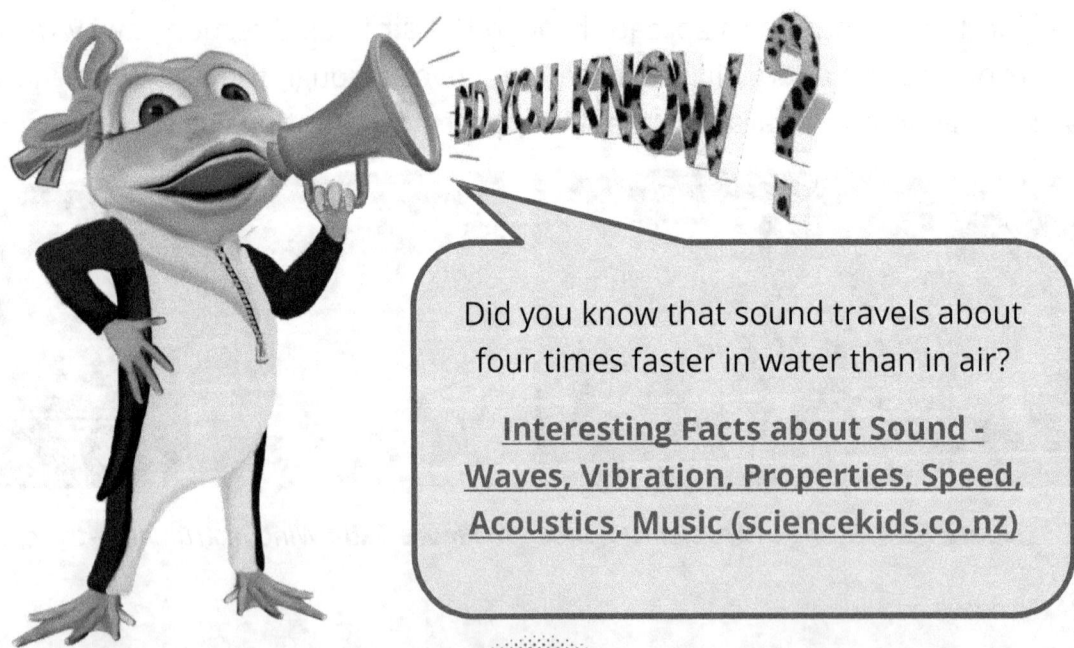

Did you know that sound travels about four times faster in water than in air?

Interesting Facts about Sound - Waves, Vibration, Properties, Speed, Acoustics, Music (sciencekids.co.nz)

Activity 5.1.1: Sounds move in all directions from the source

Work with your classmates to investigate if sounds move in all directions from the source. Listen carefully.

Materials:

- blindfolds
- radio (or anything that can make a sound)
- table

Procedure:

1. Students sit in a circle blindfolded or with their eyes closed.
2. The teacher places the radio, or the sound maker, on a table and turns it on.
3. The teacher asks students to point in the direction they believe the sound is coming from.
4. Students remove the blindfold or open their eyes and look in the direction where they pointed to see if they were correct in locating the source of the sound.

What conclusion can you draw from this activity?

5.2: Properties of sound vibrations

Have you ever heard someone scream or whisper? Touch your voice box and scream. Did you notice the changes in your voice box? Now whisper to yourself. Did you notice any difference in your voice box? If yes, what are they?

The movement of the voice box is what we call frequency or pitch.

Pitch (frequency) of sound

Sounds can also be described by their **pitch** or how low or high they seem to the listener. The pitch (frequency) of a sound depends on the number of vibrations given off per second.

Activity 5.2.1: Investigating pitch

Work in groups of three to complete the following activities relating to the frequency of a sound wave.

Part A

Material:

- one ruler (30 cm)

Procedure:

1. Place a 30 cm plastic ruler on your desk with the end extending 25 cm off the edge.

2. Allow one student to use one hand to hold a ruler firmly in place on the table with one end extending about 25 cm over the edge. The student uses the other hand to pluck the end of the ruler extending off the desk, while another student observes and records the speed of the vibration (the up and down motion of the ruler) and the frequency of the vibration using their ears.

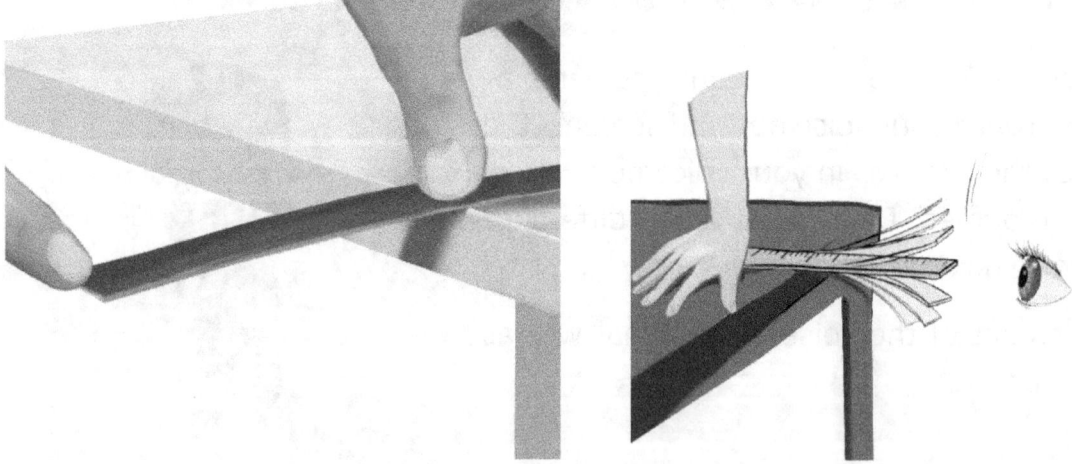

Plucking and observing the vibrating ruler.

3. Change the distance the ruler extends off the edge of the desk to 20 cm and repeat step 2 using the same amount of force.

4. Change the distance the ruler extends off the edge of the desk to 15 cm and repeat step 2 using the same amount of force.

5. Change the distance the ruler extends off the edge of the desk to 10 cm and repeat step 2 using the same amount of force.

6. Record your results in the table below. Place a tick (✓) in the box for each distance that you tested.

Distance of extension	Estimated speed of vibrations			Pitch of sound		
	Fast	Medium	Slow	High	Medium	Low
25 cm						
20 cm						
15 cm						
10 cm						

Note: *The speed of vibration depends on how quickly the ruler is going up and down. Low pitch is like when you turn up the bass on the radio. The high pitch is the 'squeaky' sound.*

Activity 5.2.2: Investigating pitch in a bottle

Next, your group will investigate differences in pitch by making a water xylophone.

Materials:

- six glass bottles of similar size and shape
- water
- permanent marker
- wooden ruler or wooden spoon

Procedure:

1. Set six empty glass bottles in a straight line.
2. Fill the first one almost to the top with water.
3. Pour a little less water into the second bottle.
4. Pour less water into the third bottle than you poured into the second.
5. Pour less water into the fourth bottle than you poured into the third.
6. Pour less water into the fifth bottle than you poured into the fourth. Leave the sixth bottle empty.
7. Number the bottles from one to six, lined up according to the height of the water in the bottles, as illustrated in the diagram that follows. That is, number one will be empty and number six almost full.

Creating a sound in a bottle

1. Blow across the top of each bottle.
2. Gently tap each bottle.

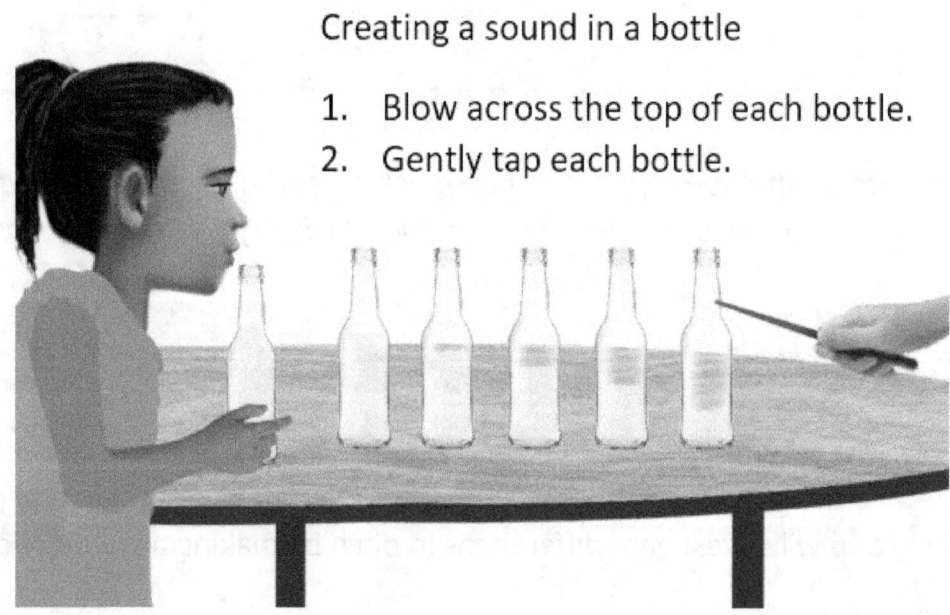

8. Gently tap the edge of each bottle with a wooden ruler or spoon using the same force.

    a. What do you notice about the sounds made as you tap from the empty bottle to the full bottle (from one to six)?

    b. How does the pitch vary?

9. Each person should gently blow across the top edge of each bottle.

    a. What do you notice about the sounds made as you blow across the edge?

    b. How does the pitch vary?

10. Compare any difference in pitch when you blow across or tap the edge of the bottle. Use your observations to answer the questions below.

    a. Which bottle produces the highest pitch when it is struck? Explain why.
    b. Which bottle produces the lowest pitch when struck? Explain why.
    c. Which bottle produces the highest pitch when you blow across it? Explain why.
    d. Which bottle produces the lowest pitch when you blow across it? Explain why.

11. Note the following true statements.
    a. When you blow across the bottle, the air in the bottle vibrates.
    b. When you tap the bottle, the water vibrates.
    c. The more air is blown, the lower the pitch.
    d. The less air blown, the higher the pitch.
    e. When you tap the bottle, the one with less water has a higher pitch.
    f. When you tap the bottle, the one with more water has a lower pitch.

12. Work together to try to play a tune.

Activity 5.2.3: Science, technology, engineering, arts and mathematics (STEAM) and me: Create your own musical instrument

Remember to use the engineering design process, as shown:

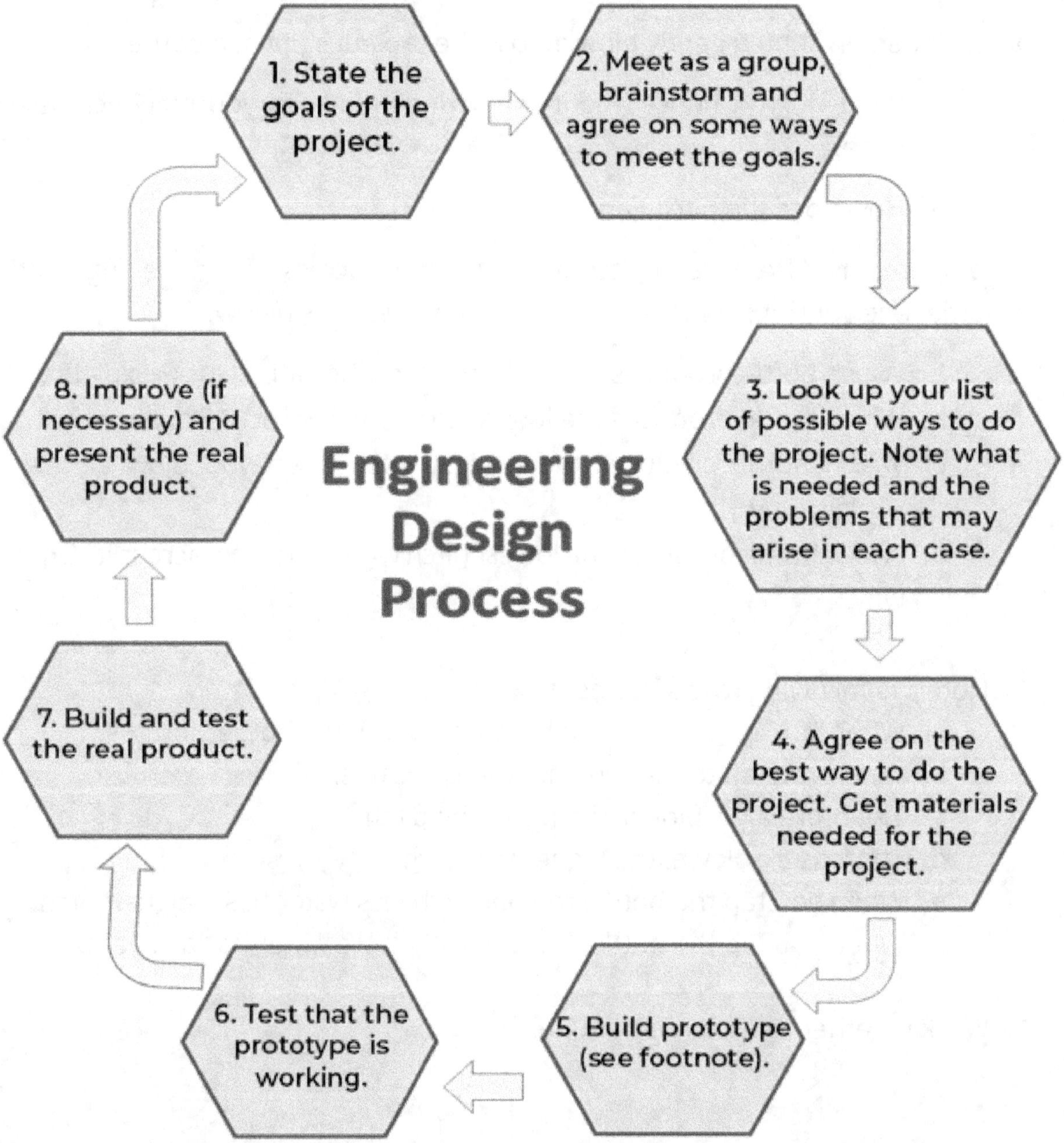

Engineering Design Process

1. State the goals of the project.
2. Meet as a group, brainstorm and agree on some ways to meet the goals.
3. Look up your list of possible ways to do the project. Note what is needed and the problems that may arise in each case.
4. Agree on the best way to do the project. Get materials needed for the project.
5. Build prototype (see footnote).
6. Test that the prototype is working.
7. Build and test the real product.
8. Improve (if necessary) and present the real product.

Note: Step 5. Prototypes or trial versions may be built and tested using relatively inexpensive materials to avoid wasting resources. For example, using paper instead of cardboard or hardwood or using glue instead of cement to reduce expense. Step 6. If the prototype tested does not work as it should, you may need to go back to the brainstorming step (#2).

CHEETAH® PREPARING THE JAMAICAN SCIENTIST

Background: There is a shortage of musical instruments within your school. Students who are interested in music are asked to get together in groups to make instruments for the upcoming festival.

Material: Based on the instrument you choose, select your own material.

Procedure:

1. Get together in groups and brainstorm possible designs for your instrument.
2. The team should draw a labelled diagram showing the design of the instrument. This should include a labelled diagram of the instrument and a list of materials needed to make it.
3. Write step-by-step instructions for making the instrument. This should include:
   - how the instrument will produce sound
   - how the loudness of the sound produced by the instrument can be changed
   - how the pitch of the sound produced by the instrument can be changed.
4. Share your plans with your teacher.
5. Construct the instrument, making sure that all the members of the team are involved.
6. Use the instrument to investigate and answer the following questions:
   - State the materials used to build the instrument and why they were selected.
   - How does the instrument produce sound vibrations?
   - How do the dimensions (length, thickness and shape) and the other design features of the instrument create and change the pitch?

Let's review the properties of sound vibration.

1. What properties of a sound vibration determine a sound's volume?
2. What properties of a sound vibration determine a sound's pitch?
3. Why does a sound need a medium to move from the source to the ears?
4. Explain why the volume of sound decreases as someone moves further away from the source.
5. Lightning gives off both light and sound at the same time. Why is the flash of light seen before the sound of thunder is heard?

5.3 Reflection of sound vibration

If you have ever sung in the shower, you may have noticed that your voice sounded louder than normal. This is sound vibrations bouncing off the wall of this small, empty room. The reflection of a sound vibration is called an **echo**.

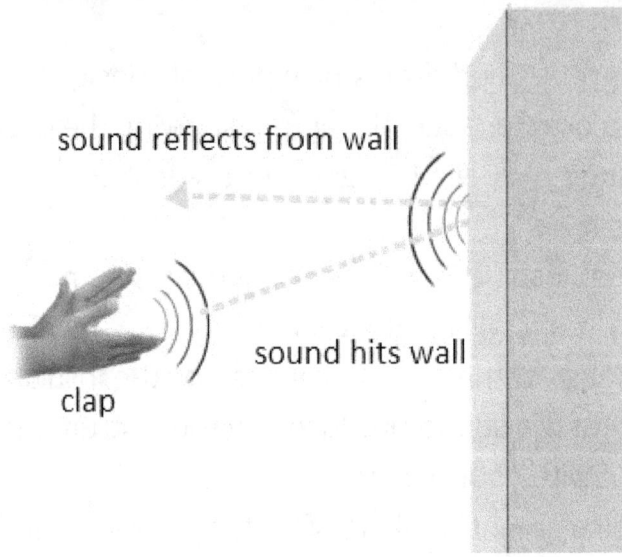

Figure 5.5: *Echoes formed by the reflection of sounds.*

Absorption of sound

Sound vibrations do not always bounce off surfaces; depending on the surface, sounds can be absorbed or muffled. This is called **sound absorption**. Materials that muffle sounds (absorbers), stop or prevent the sound from being heard. This is called **acoustic insulation**. Materials that reflect sounds allow sound to be heard.

For example, if you were to try playing basketball on a sandy beach, you would notice that the ball cannot bounce very well. This is because the sand absorbs the energy from the ball. In the same way, the energy of sound waves is absorbed by soft surfaces such as rugs, curtains, sofas and plants. This energy is reflected from hard surfaces, such as tile or concrete.

Some solids, such as glass or metal, are very good at transmitting sounds. Other solids, like carpet and fabric, muffle sounds and are used to create soundproof rooms, such as music studios.

CHEETAH® PREPARING THE JAMAICAN SCIENTIST

Activity 5.3.1: Science, technology, engineering, arts and mathematics (STEAM) and me: Test different materials for their ability to absorb sound

Scenario: You find it easier to study at home. However, your neighbour plays music and you would like to cover your window with materials that can absorb sound to reduce the volume of the sound heard in your room. You do not know to what extent different materials absorb sound movement, so you would like to run a test to select a suitable material.

Hypothesis: Predict or guess the type of material you think will best keep out the noise from the list below.

Materials:

- cartridge paper
- cardboard
- bubble wrap
- fabric
- newspaper
- foam
- shoebox
- cell phone

Procedure:

1. Working in groups of three or four students, get a box and padding materials to use for testing.
2. Using the material to make a soundproof box, line the box with one of the materials to be tested. The lining should be exactly 5 cm in thickness on all sides of the box, including the top and bottom.
3. Turn on the alarm feature on the cell phone and increase to maximum volume. Place the cell phone in the box. Close the lid (cover).
4. All the team members should listen to the sound and record their observations for that material.
5. Remove and turn off the cell phone alarm.
6. Remove the lining of the box and replace it with another material. Retest using steps 3 and 4 of this procedure. Do this for all the materials to be tested.

Follow-up

- Compare your hypothesis with your observations.
- Describe the characteristics of the materials that were best at absorbing sound.

- How did you know which material was best for absorbing sound?
- What kind of material would you recommend using to keep the sound out of the room?

Let's review the absorption of sound

1. How are echoes created?
2. How does the number of objects in a room affect the sound heard in that room?
3. How do soft surfaces, like carpets and curtains, reduce the amount of sound reflection in a theatre?

Activity 5.3.2: Information and communication technologies (ICT) and me: Sound in outer space

Does sound travel in outer space? If we were standing on the moon, would it be possible for us to have a conversation? Research to find out if this is possible.

Use what you have learned to create a cartoon or short story explaining sound in outer space. Share your creation with your classmates to explain your findings.

Did you know?
Acoustics is the study of how sound travels.

CHEETAH™ collaboration corner

Activity 5.3.3: Sound and matter

Work with a partner to investigate how sound travels through different states of matter.

Materials:

- one pencil
- desk or table

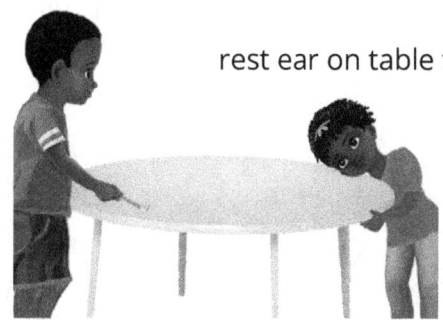

tap on table

rest ear on table to listen

Sound movement through air and solids

1. Tap on a desk, while your partner listens and records their observations.

2. Have your partner place their ear on the desk while you continue tapping, then record their observations again.

3. Switch roles and repeat the experiment.

4. Compare the differences in the sound through the desk and the air in your pair, then discuss your observations with your class.

Note: The ear resting on the desk hears the sound coming through the solid. The ear that is not resting on the desk hears the sound coming through the air. So, the listener is expected to hear two sounds and give differences between these two sounds.

Let's take another look at how sound travels through matter.

1. What are the THREE states of matter?

2. Compare the speed at which sound moves through solids and liquids.

3. What is a vacuum? Is sound able to move through a vacuum?

4. Why would the Maroons put their ears to the ground to tell if soldiers were coming?

5. Why would sound travel faster in hot water than in cold water?

5.4: Pleasant and unpleasant sounds

Pleasant sounds are those that people like to hear, while unpleasant sounds are considered to be noise. Whether a sound is pleasant or unpleasant may be based on an individual. For example, a man playing music in his backyard may believe the music is pleasant. However, a student studying nearby may consider this unpleasant.

What is noise?

Whenever the loudness (amplitude) of a sound is unwanted or unpleasant, it is called **noise**.

Figure 5.6: Blaring sounds cause noise.

Noise pollution and the sources

Noise pollution occurs when sounds become too loud, annoying or harmful to human hearing. Sources of noise pollution include firecrackers, jet airplanes, factories, automobiles and traffic, high-powered equipment, motor vehicle horns, sirens, loud music and roadwork compressors.

Figure 5.7: Man using a jackhammer to break apart hardened concrete.

Harmful effects of noise pollution

Noise pollution can cause hearing loss in humans. Our ears are organs with extremely small, delicate structures that must work together for a person to have normal hearing. If any of these parts inside the ear are injured, we may have hearing loss.

Noise pollution also affects marine animals such as whales and dolphins that use sound for communication, navigation or finding food. The noise from ships and industrial activities near the ocean (such as oil factories) creates background noise in the water. This makes it impossible for animals to communicate.

Other effects of noise pollution include:
- sleep disturbances or disorders
- trouble communicating
- increased aggression in pets and people
- headaches
- nausea.

Jamaica passed a law in 1997 called The Noise Abatement Act, which restricts making loud sounds within the community. Here is a section from the law:

> No person shall, on any private premises or in any public place at any time of day or night- (U) sing, or sound or play upon any musical or noisy instrument; or (b) operate, or permit or cause to be operated, any loudspeaker, microphone or any other device for the amplification of sound, in such a manner that the sound is audible beyond a distance of one hundred metres from the source of such sound and is reasonably capable of causing annoyance to persons in the vicinity so, however, that where during the period specified in subsection (4) such sound is audible beyond that distance in the vicinity of any dwelling house, hospital...nursing home, infirmary, hotel or guest house, such sound shall be presumed to cause annoyance to persons in that vicinity.

Activity 5.4.1: Concern for the school community

After reading the law, Shantell realised that there was no mention of schools. She discussed this with her friend, Tony, and they both decided to write a letter to the Prime Minister asking why schools were not mentioned in the law.

Support Shantell and Tony by writing a letter to the editor outlining your concerns, and requesting that schools be given special mention.

Preventing hearing loss

Hearing loss is one major effect of noise pollution. Fortunately, several methods can reduce the effects of noise pollution.

These include:

- turning down devices that produce sound
- wearing hearing protection such as earplugs and earmuffs
- shutting doors and windows that open to noisy areas
- using double-glazed doors and windows on buildings near noisy areas
- using soundproof rooms for noisy machinery.

Figure 5.8: *Earmuffs used to reduce sound going to the ear.*

Let's examine the causes and effects of unpleasant sounds.

1. What is noise?

2. Why is it important for people to stay away from loud sounds?

3. At what decibel level does hearing loss become a concern?

4. Make THREE recommendations for preventing hearing loss.

5. Write a statement giving your opinion about loud music on public passenger vehicles in Jamaica.

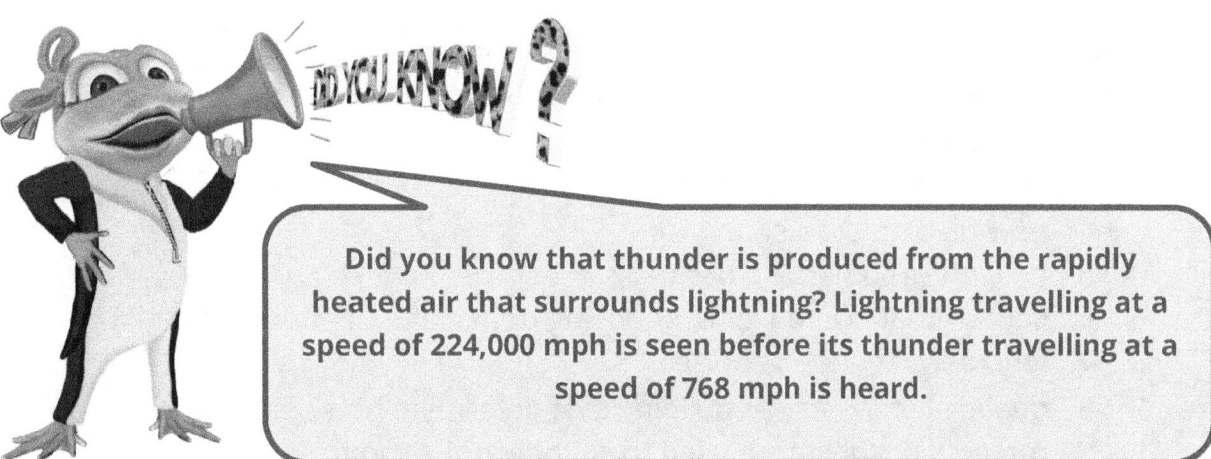

Did you know that thunder is produced from the rapidly heated air that surrounds lightning? Lightning travelling at a speed of 224,000 mph is seen before its thunder travelling at a speed of 768 mph is heard.

Activity 5.4.2: Sound unscramble puzzle

Unscramble the following using the clues below.

1. nosdu
2. itnraviob
3. eolmuv
4. chitp
5. necfyequr
6. seoni
7. ptnoliolu
8. cocuiasti
9. tinanosiul

Clues

1. **vibrations** moving through air or water that can be heard
2. small quick movements back and forth
3. how high or low a sound is
4. how high or low a sound is
5. how often something occurs
6. sounds that are too loud, annoying or harmful to humans or animals
7. loud and repetitious noise
8. qualities of a room or building that determine how sound moves in it
9. material or substance used to soundproof a room

Evaluate yourself!

Use this evaluation grid to check your understanding of the concepts discussed in this chapter. Read each statement below and insert the symbol that best shows how well you feel you understand the concept. Ask a teacher or parent to help you go over any areas that are still unclear, or that you do not feel you have mastered. Be honest!

I got it!

I need to do more work.

I do not get it. I need help.

In this chapter:

	I got it!	I need to do more work.	I do not get it. I need help.
1. I can identify the effects of loud sounds.			
2. I can explain why sounds may be thought of as pleasant or unpleasant.			
3. I can recognise sources of noise pollution and list ways to eliminate them.			
4. I can describe sounds using appropriate scientific language.			
5. I can create hypotheses when doing investigations into the properties of sound.			
6. I value individual effort and teamwork and I work cooperatively in groups.			

TERM 2, UNIT 1:
Materials – Properties and uses

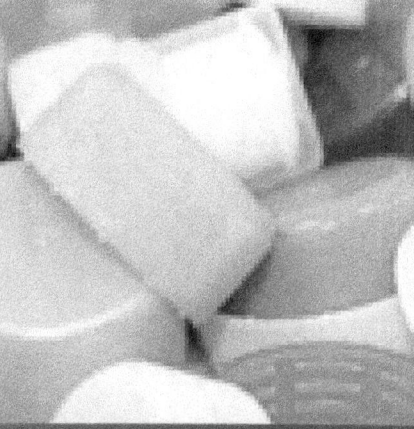

MATERIALS

Words worth knowing

The list below contains the scientific words that are used in this chapter.

- absorbency
- condensation
- conductivity
- elasticity
- evaporation
- freezing
- gas
- hardness
- irreversible change
- liquid
- magnetism
- melting
- plasticity
- properties
- reversible change
- solid
- strength
- transparency
- water resistance

Whether it rusts, resists, absorbs, or reflects, materials have secrets, and it is your job to uncover them! Are you ready? Let's go!

Chapter 6: Everyday materials

Chapter objectives

- ✓ Examine a selection of materials/objects to determine the transparency, absorbency, strength, magnetic property, and heat conductivity of materials in everyday use.
- ✓ List some properties of materials that determine the choice of objects for a specific purpose in everyday life.
- ✓ Identify correct and safe ways of using, storing and disposing of materials and household items.
- ✓ Evaluate how the disposal of selected materials affects the environment.
- ✓ Assess the impact of different materials on society.
- ✓ Generate predictions of material properties based on observations and experience.

CHEETAH Science Fiction

Everyday materials

Boats

Croaky Croak, the frog, was in a bad mood. The boat she had built was sinking in the river, but she couldn't understand why. It had no holes in it and it certainly wasn't heavy. Hooty Hoot chuckled to herself in a tree but decided to fly down and tell Croaky Croak what was wrong.

'You made the boat out of cardboard,' she said. 'And cardboard isn't waterproof. That's why it is sinking in the river.'

'Thank you for telling me,' said the friendly frog. 'I'll go home and make another boat. But this time I won't use cardboard.'

The next day, Hooty Hoot saw Croaky Croak on the riverbank. She was crying and pointing to the water.

'What's wrong now?' asked the owl.

'I made another boat,' she said. 'This time I used house bricks because I knew water wouldn't ruin them like it did the cardboard. But it sank.'

Hooty Hoot shook her head from side to side. 'House bricks are too heavy,' she said. 'You need to use a material that floats. Why don't you reuse plastic bottles to make a boat?'

The next day, as Hooty Hoot was flying to the market, she spotted Croaky Croak waving to her from the river. She was playing with her boat in the water, but still looked a little sad.

'At last,' said Hooty Hoot. 'You've made a boat that floats! But why the long face?'

'Because there's so much STUFF!' said Croaky Croak. 'Why can't everything be made of the same material?'

Hooty Hoot explained that all materials have different uses, and one material cannot do everything. 'Metal trousers would be very uncomfortable,' she chuckled. 'And who would use a glass hammer to bang in a nail? Or wipe the windows with a shovel?'

Croaky laughed, 'Okay, okay. I get it! Come, let's go for a ride in my new boat.'

6.1: Methods for selecting materials

Properties of materials for use

Materials are used to make different things (objects). Different types of materials are used to make different things. Some things are made from more than one type of material. For example, a chair can be made from metal, plastic or wood. The choice of materials used to make an object will depend on the properties of the material and the use of the object. For example, a wooden spoon is the best choice to stir a pot of soup, although a spoon can be made from metal or wood. These properties are very important to help determine how useful a material might be for a certain project. For example, if you wanted to build an outdoor wall, you would not choose paper as your material.

The purpose of the task helps to determine the best material for a job, as shown in the picture.

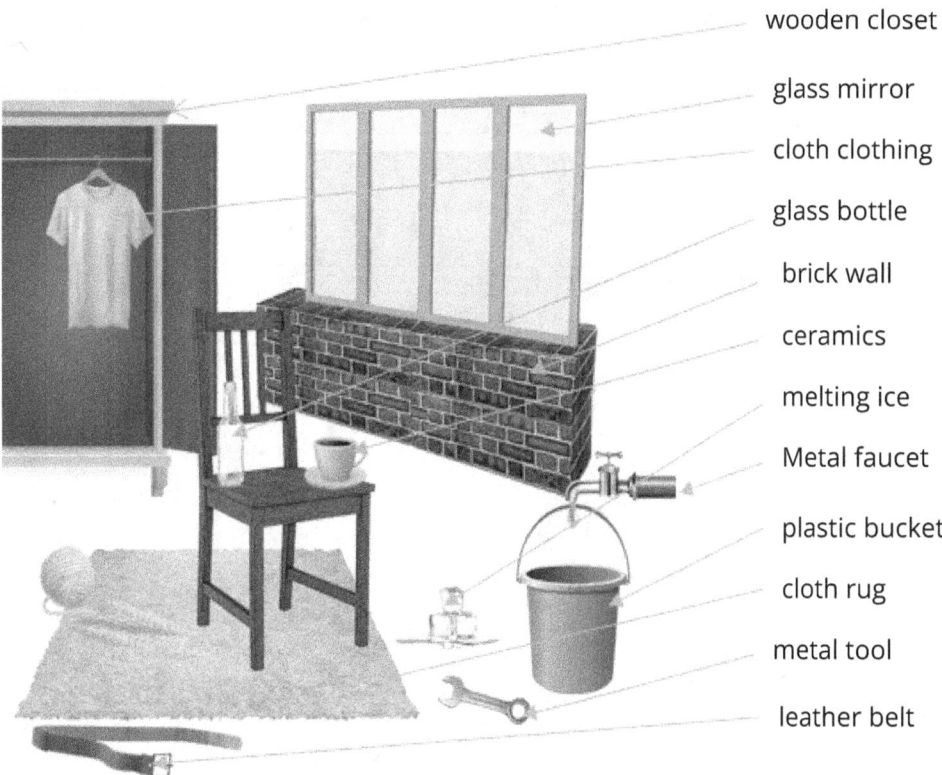

Figure 6.1: *Materials used to make different things (objects).*

When choosing materials to make an object, the following properties of each material should be considered:

1. transparency
2. absorbency
3. strength
4. magnetism
5. heat conductivity

Transparency

Transparency is the ability of a material to allow light to pass through or to be easily seen through. For example, clear glass and plastic are transparent.

Activity 6.1.1 Transparency of material

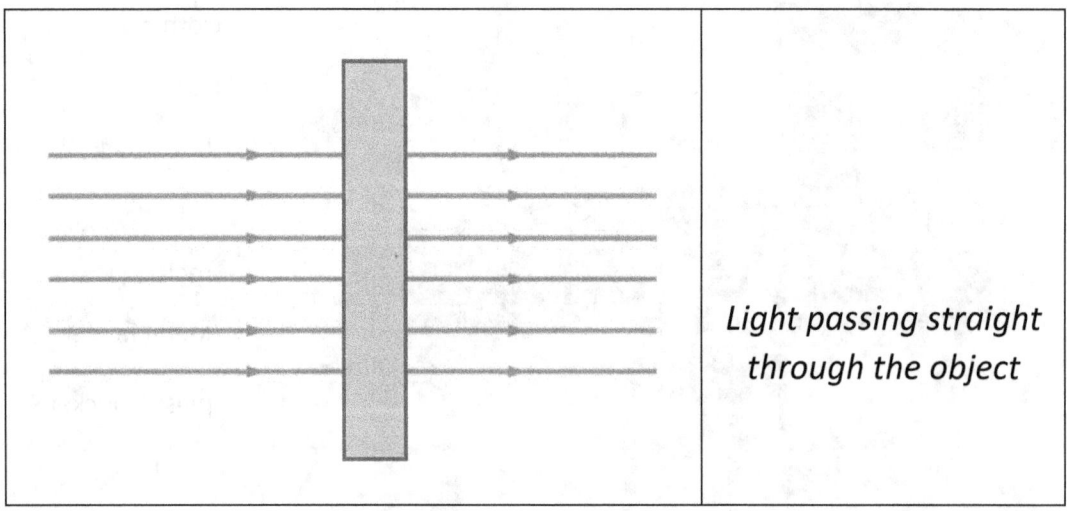

Light passing straight through the object

Materials:

- clear plastic cup
- not so clear plastic cup
- coloured plastic cup
- water
- blue pencil

Procedure:

1. Set out the labels of the three small plastic cups.
2. Add three-quarters water as shown in the drawing.

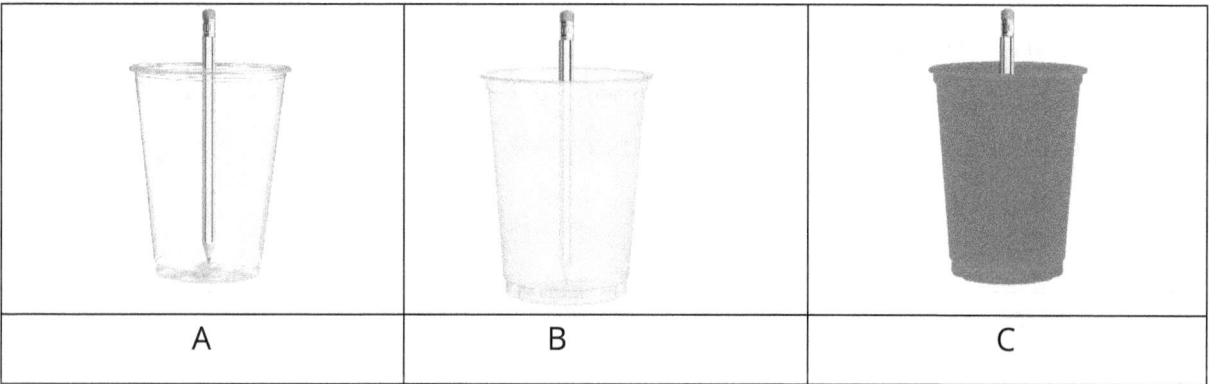

3. Insert a blue pencil into each and note observations of the pencil from outside.

Use what you know and your observations to answer the following questions:

1. Which cup is easiest to see through?
2. Which cup is made from the type of material that is best suited to block out the sun?
3. Why is there a difference in how the part of the pencil in the cup is viewed?
4. Which cup allows the least light to pass through?
5. Which cup would be least likely to cast a shadow?

Absorbency

Absorbency is the ability of materials to take in or 'soak up' liquid. For example, sponges, cloth and some types of paper are absorbent, while plastics, wood and metals are not. The opposite of water absorbency is water resistance. Water resistance describes a material's ability to repel or hold off the water. Metals and plastics are very water resistant.

Activity 6.1.2   Testing the absorbency of different types of paper

Materials:

• water • transparent container • stopwatch • one drinking straw • tape	• 2 cm x 10 cm strips of different types of paper (tissue, hand towel, newspaper, cartridge paper, grease paper, book-leaf, certificate paper and/or cardboard) • permanent marker

Procedure:

1. Pour water into a transparent container, as shown in the diagram.

2. Mark the starting level on each paper strip, 1 cm from the tip.

3. Tape the strips from each sample of paper to the drinking straw, ensuring that each strip is the same length when hanging from the straw as shown in the diagram.

4. Carefully place the drinking straw across the mouth of the transparent container hanging the strip down so the tips of the papers are submerged in the water up to the 1 cm mark. **(Note:** *Ensure that the strips of paper touch the water at the same time.)*

5. Allow the paper to hang in the water for 60 seconds, remove them and quickly mark the spot where the watermark ends.

straw

line showing where water reaches up the paper after 60 seconds of absorption

line showing where water meets paper at the start of absorption

Use your observations to answer the following questions:

1. Calculate the speed (in cm per minute) at which each paper absorbed water.
   Speed = distance moved by the water ÷ time in minutes
2. Why is it important to test absorbency in a different material?
3. Why was it important that all the strips touch the water at the same time?
4. How do you know which paper was most absorbent?
5. List the papers in order of greatest water resistance.

Strength

Strength is the ability of a material to not break, tear, stretch or bend when a great force is added to it. For example, the strength of a piece of wood is its ability to not bend when a force is applied to it (when you hit it). Different materials have different levels of strength.

Figure 6.2: Building made of wood.

Figure 6.3: Building made of concrete.

For example, metal rulers are harder to break than wooden rulers. Hardness is a measure of strength. It is how well an object can resist scratching and pressure. For example, hardwood does not get scratched as easily as softwood. This information is very important for the construction of furniture in a large building. A concrete building can be much larger than a wooden building because concrete is much stronger than wood.

Did you know that a blade made from the volcanic glass obsidian is 500 times sharper than a surgeon's steel scalpel?

Activity 6.1.3 Testing material strength

Test the strength of materials using each of the methods below.

Materials:

- a small plastic bag
- 50 marbles
- cloth
- aluminum foil
- paper towel
- ruler
- scissors
- masking tape

- newspaper
- plain paper
- plastic bag
- cartridge paper
- two wooden blocks or supports

Procedure: The bending method

1. Working in groups of three, cut each of the materials into 30 cm x 35 cm sized strips.

2. Roll each material into a tight roll 35 cm long (with equal tightness).

3. Secure the ends with tape.

4. Place each roll of material to be tested across the wooden blocks as shown in the diagram, leaving the space between the blocks at 30 cm.

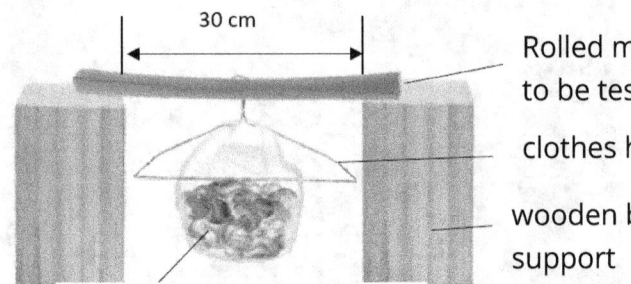

5. Begin by placing two marbles in the bag that is attached to the hook.

6. Hang the hook from the centre of the rolled strip to be tested.

7. Keep adding two marbles at a time until the strip bends.

8. Record the total number of marbles that cause each material to bend.

Type of Material	Number of marbles needed for bending

Use what you learn from the experiment to answer the following questions:

1. Why were the materials cut to the same length and folded in the same way?
2. Why did your material bend when the weights were added?
3. Which material bends the easiest?
4. Which material took the greatest force to bend?
5. Which material would you recommend to make a walking stick?
6. What are the strongest and weakest materials?

Examine the buildings below, then use your observations to answer the questions that follow:

1. What materials were used to build house A?
2. What materials were used to build house B?
3. Which building would be better for temporary storage? Explain your answer.
4. Which building would be able to withstand stormy weather? Explain your answer.
5. Why do you think it is important to consider the location and purpose of a building when choosing the materials for building it?

List the types of materials used to build houses in Jamaica and describe why each material is effective in this country.

Materials	Reasons for choosing the material

Magnetism

Magnetism is the ability of a material to be attracted to a magnet. A magnet is a piece of metal or rock that has the power to pull some metals toward it. **Magnetism** exists as a force in nature and does not have to be touching each other to work.

Materials like plastic or glass are not magnetic, while metals such as iron and nickel are highly magnetic. The two ends of the magnet are called its poles. The poles are named the north magnetic pole and the south magnetic pole. Opposite poles of a magnet attract each other (north pole—south pole). The like poles of a magnet repel each other (north pole—north pole or south pole—south pole).

Magnets are not attracted to all metals such as aluminum and copper.

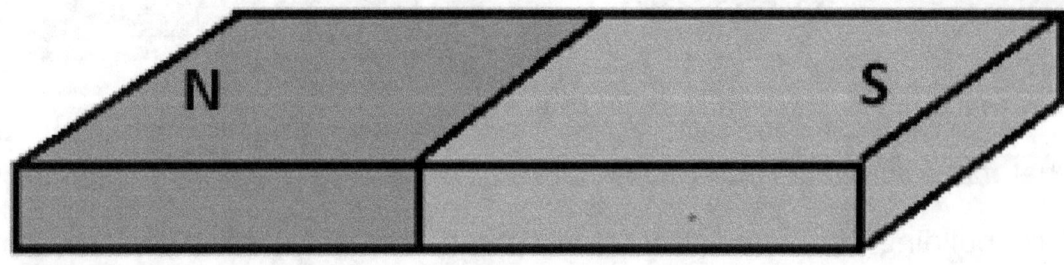

Figure 6.4: *Magnet showing north and south poles.*

Activity 6.1.4 Testing magnetism

Complete this activity individually or in suitable groups as selected by your teacher.

Materials:

• bar magnet (any kind of magnet) • paper • cloth • different types of nails • cardboard • different types of keys • wood	• stone • plastic • aluminum (foil paper) • copper • paper clip • pen • different types of coins

Procedure:

1. Place a bar magnet over each of the different objects you collected.

2. Determine which of the objects are attracted to the bar magnet. These objects have magnetic properties.

3. Create a table showing which of the items you tested were magnetic and which items were non-magnetic.

Types of material	Magnetic (tick)	Non-magnetic (tick)

4. Are all metals magnetic?

Heat conductivity

Have you ever left your metal spoon in your cup of hot tea and the handle of the spoon becomes hot? Do you think it would be the same with a wooden spoon? Heat conductivity is the ability of a material to allow heat to pass through it. Metals are good conductors of heat, while non-metals are not good conductors of heat and are called insulators.

Activity 6.1.5 Testing heat conductivity of materials

Perform this activity in a group of four persons.

ATTENTION
Adult supervision required for this activity.

Materials:

- beaker/ container
- hot water
- small plastic spoon (scoop)
- plastic tablespoon
- metal tablespoon
- wooden tablespoon
- soft butter/margarine

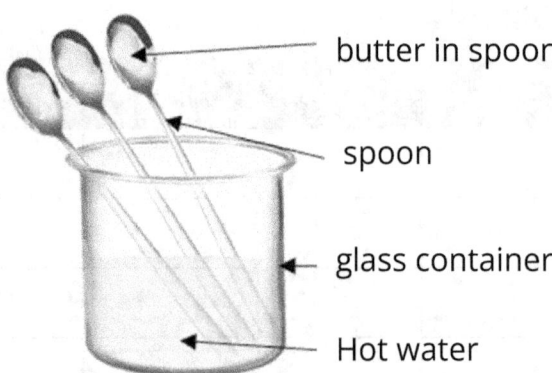

Figure 6.5: Testing heat conductivity of three types of materials.

Procedure:

1. Get three different tablespoons made from wood, plastic and metal.

2. Use a small plastic spoon to scoop out one level spoonful of butter/margarine and pack it into each of the spoons to be tested. Ensure that when the spoon is slanted downwards, the butter sticks to the spoon.

3. The teacher should half-fill the beaker with hot water.

4. Place the handle of each of the spoons into the hot water, so that the handle rests on the edge of the beaker and the neck of the spoon does not touch the beaker as shown in the diagram.

5. Allow the spoon handles to sit in the hot water for about two minutes. Record your observations.

CHEETAH™ collaboration corner

In groups of three, complete the following activities:

Use your observations to answer the following questions:

1. In which spoon did the butter melt first?
2. Why did the heat cause the butter to melt?
3. List the spoons according to their ability to conduct heat.
4. Which material would you use to make the handle of a cooking pot? Explain your answer.
5. Poor conductors of heat are insulators. From the experiment above, what type of material is the best insulator?

Did you know?
Heat will flow from one point to another if there is a difference in the temperature between the two points.

6.2: Safe use of household items

The use of product labels

Many household products (such as paints, cleaners, batteries, adhesives and pesticides) contain poisonous chemicals, which can be dangerous to health and the environment. All products have labels that ensure we are safely using them. These labels make buyers aware of the contents, expiration dates, storage and disposal instructions and safety warnings. Labels must be read and safety precautions followed when using household products. This helps to reduce health problems, fires and pollution.

Figure 6.6: Household cleaners.

Using products safely

Here are some general tips on how to use household items and other household materials safely:

1. Always read the label before you use a product.
2. Always follow the directions for use and heed safety warnings.
3. Only use products for the purpose listed on the label.
4. Open windows and doors when you are using products with potentially harmful chemicals, such as cleaning agents and paints. Use fans to blow the odour away from the work area and towards an open window or door.

Figure 6.7: Precautions on the product label.

5. Wear protective gear, such as masks, aprons, gloves and glasses or goggles, if you deem it important or as the label instructs.

6. Do not mix products together. This can produce toxic fumes that can be harmful to your health and the environment.

7. Do not use certain household products around pregnant women, sick people or young children. Some products may cause an allergic reaction when inhaled or in contact with the skin. Seek medical care if you develop an allergic reaction.

8. Move away from work areas when eating. If you are unable to move to another area, make sure to keep your food covered.

9. Wash your hands after using household products or any material that contains potentially hazardous chemicals.

10. Seek medical care if you have inhaled or ingested household cleaners, where the label warns against it.

Label features

Most labels on household products and other materials contain the following information:

- *Manufacturer*: The name and contact information of the company that made the product.

- *Expiration date or expiry date*: The date at which the product or material may no longer be useful.

- *Manufactured date*: The date when the product or material was produced and packaged.

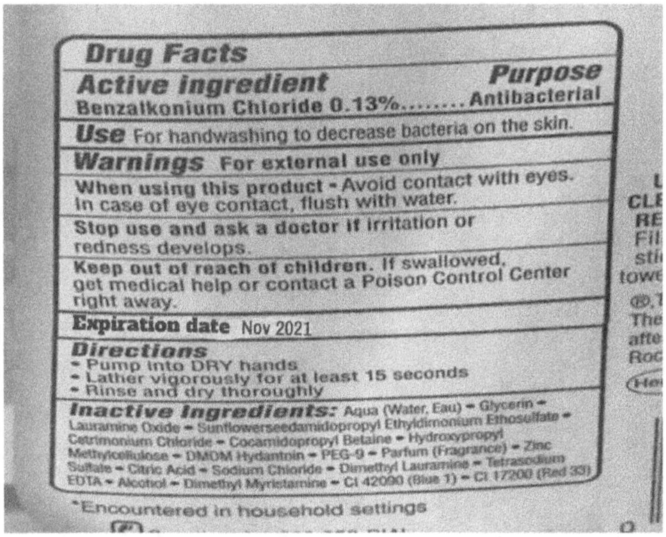

- *Product batch number*: Some products are made in groups and packaged in separate containers. These products are described as a batch. When one product in a batch is defective, all

Figure 6.8: *Drug facts on the product label.*

the products in that batch may be defective. The batch number allows a manufacturer to track and recall all the products in a batch if they are defective.

- *Ingredients*: A list of the specific resources that were used to make the product or material.

- *Instructions or directions*: A description of how to use the product or material properly. For example, instructions relating to children—keep out of the reach of children.

- *Barcode*: A unique identifier that provides information about the product or material. Barcodes are read by computer scanners.

Figure 6.9: *Barcode on product label.*

- *Storage*: A list of instructions for how and under what conditions the product or material should be stored, for example, keep in a cool, dry place.

- *Disposal*: A list of instructions that explain how to dispose of the product or material and its container after it has been used or has expired.

Figure 6.10: *Storage and disposal on the product label.*

- *Warning signs or messages*: These explain how the product or material can be safely used. This feature may not be found on all items.

- *First aid information*: This is used if the product is used inappropriately or causes an injury.

Figure 6.11: *Caution or warning instructions on the product label.*

- *Manufacturer's contact information*: If there is an emergency or the product is recalled.

Warning labels

There are many different types of *warning symbols* used on household items to warn individuals of the danger of the contents and to advise users on how to use them properly. Here are some safety warning symbols.

Hazard symbols

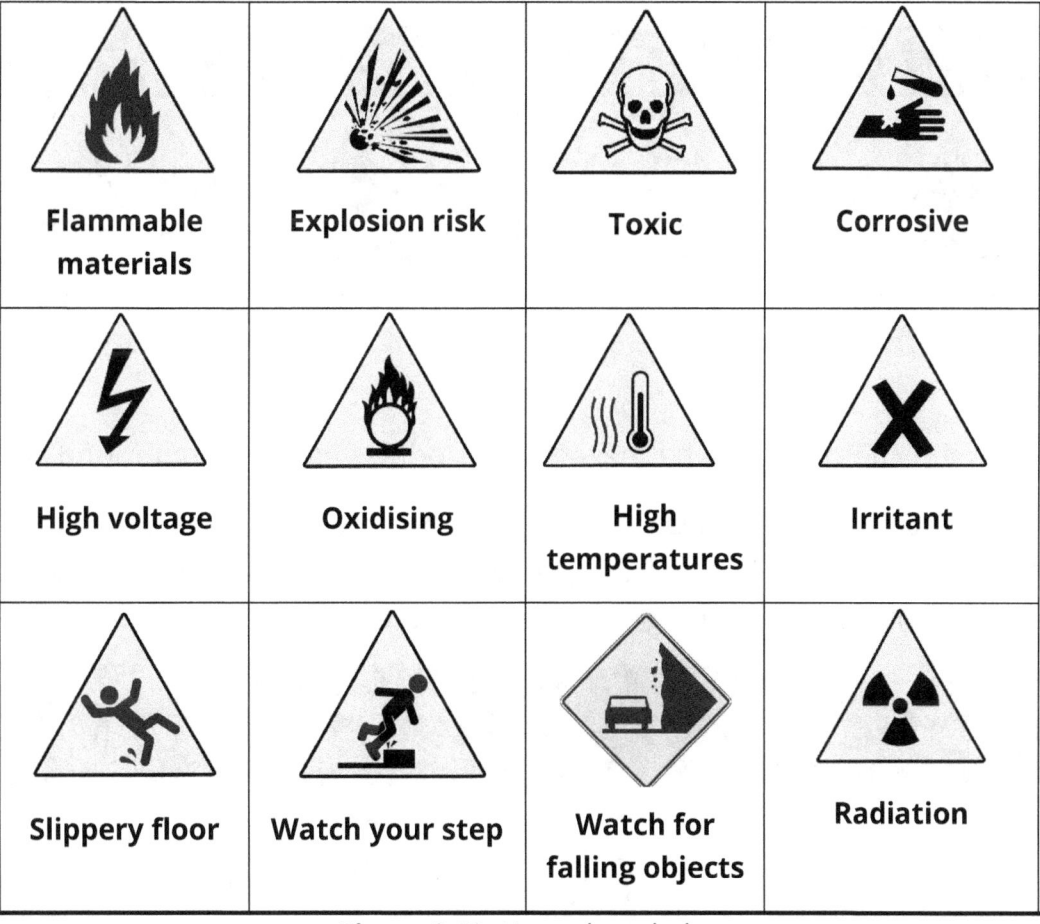

Figure 6.12: *Hazard symbols.*

Let's take another look at warning labels.

1. Which of these hazard labels relate directly to our homes?
2. What are THREE reasons warning labels are important?
3. What are THREE types of information you can find on a household product label?
4. Which TWO signs would be important in the kitchen?
5. Which sign could be found on the bottle of a drain cleaning solution?

Activity 6.2.1: Identifying warning labels

Part 1:

With the assistance of your parents, look around your home for products with the following symbols shown below. Record the name of the product and its purpose and the symbol that was on it in your notebook. Share your findings with your classmates.

A	B	C	D
⚠️ (explosive)	⚠️ (skull and crossbones)	⚠️ (flame)	⚠️ (corrosive)

Part 2:

In groups of THREE, discuss what hazardous products each student found at home.

As a group, create a table similar to the one below by collating the information from each group member and listing the possible hazardous products they have at home. Provide a drawing of the warning symbols the products contain.

Hazardous household products	Draw symbols that show hazardous material	Highly/Moderately hazardous
Example: Insect spray	⚠️ (skull and crossbones)	Highly hazardous

- Work together to group the hazardous products that you found as:
    - Highly hazardous (danger/harmful, poison)
    - Moderately hazardous (caution or warning)
- Discuss with your class where your household stores these products and if this storage space is safe. Why or why not?

Part 3:

1. Research the following labels:
   - fungicide
   - insecticide
   - herbicide
   - weedicide.

   Hint: https://www.syngenta-us.com/current-label/switch_62.5wg

2. With the help of a parent/guardian, find bottles around the home with labels. Look for safe-use labels on the bottle. Take a picture of the labels, and put the safety precautions and side effects and safe disposal methods below each example given. Make an electronic portfolio with your labels.
3. Research the weed killer called Gramoxone and create a brochure informing farmers on how to use it safely.
4. Research news articles on the causes of exploding mobile phone batteries. Share with your class and teacher. Create a poster showing the safe use and disposal of batteries.

Storage of household items

Most labels give clear instructions on how to store household products. If you follow the instructions, you will not risk your health or encounter environmental problems. Here are some ways that you can properly store household products:

- Place chemicals on the highest shelf of the storage area, away from children.
- Store household items or materials in their original container with readable labels.
- Store flammable chemicals and batteries in areas that are not directly in contact with sunlight.

Did you know that graphene is the world's strongest material? It is at least 200 times stronger than steel. Graphene is used to make durable lightweight sportswear, surgical gloves and in smartphones and other equipment.

Graphene Uses (graphene-uses.com)

Activity 6.2.2: Danger in the kitchen

Read the story below.

> Gabrielle's mother bought a kitchen cleaning product that she saw advertised on TV. Unfortunately, she did not read the warning label before using the product. So, she didn't wear gloves while she was cleaning. She left the bottle open, which caused her eyes to water. Afterwards, she noticed that the skin on her hands began to peel.
>
> One day, she accidentally left the product on the kitchen counter near the stove, while she was cooking dinner. From the other room, she heard an explosion. When she ran back into the kitchen, the stove was on fire! Fortunately, she had a fire extinguisher that she was able to use to put out the fire. She later discovered that the new cleaning product caused the fire. She looked at the bottle and saw the explosive and corrosive symbols on the partially destroyed label. She was very alarmed when she realised how harmful her mistake could have been.
>
>

Use the information from the story to answer the following questions:
1. How could Gabrielle's mother have prevented damage to her hands and eyes?
2. Do all household chemicals cause fires?

3. How could Gabrielle's mother have prevented the fire?
4. Do you think she could have prevented the fire if she had read the label? Why or why not?
5. Why is it important to read labels on household products?
6. Why shouldn't you buy products without labels?

6.3: Disposal of household products

Figure 6.13: *Household cleaning products.*

Household products may contain harmful and poisonous materials, especially if they are mixed with drinking water and the soil. Here are some tips to ensure household products and other harmful materials are disposed of appropriately:

- *Household cleaning products:*

Follow the disposal instructions on the containers of household products. Containers without warning labels could be recycled or reused for school projects. If they cannot be reused or recycled, discard them responsibly.

- Farm or garden chemicals (e.g. fertilisers and pesticides):

Some farmers dispose of farm chemicals incorrectly by washing excess chemicals from the containers into rivers or in garbage heaps on their land. There is also the practice of overusing chemicals that leach into the soil and underground water storage supply (aquifer).

Empty containers of farm or garden chemicals should be wrapped in newspaper and placed in a garbage bag, then taken to the hazardous waste collection site of the National Solid Waste Management Authority (NSWMA).

Figure 6.14: *Farmer sprays insecticide over crops to reduce pests.*

- *Used automotive products (motor oil and oil filters):*

Engine oil should be recycled appropriately by an auto mechanic, not thrown away into the environment or dumped into gullies. Improperly disposed waste oil can end up in the rivers and seas and form tar balls on the beach. This can damage our marine life, make our beaches unsightly and destroy the tourist industry. Used automotive products should be placed in a temporary storage area until they can be taken to a designated dumping area at the NSWMA or to the proper recycling facility.

- *Batteries:*

Batteries should not be disposed of in a fire, as this will cause them to explode. Some batteries can be recycled.

Car batteries are high in lead, which is a very harmful chemical and can affect the human brain. For example, when a car battery is thrown away into the gully, the lead in the battery ends up in waterways. Animals such as small fish eat the lead, bigger fish eat those small fish and humans eat the big fish.

Figure 6.15: *Batteries*

Therefore, humans end up with unsafe levels of lead in their bodies and this can damage our brains, making learning impossible and causing diseases such as Alzheimer's.

- *Recyclable plastics, PETE and HDPE containers and other materials:*

Any container that bears the symbols shown to the right can be recycled.

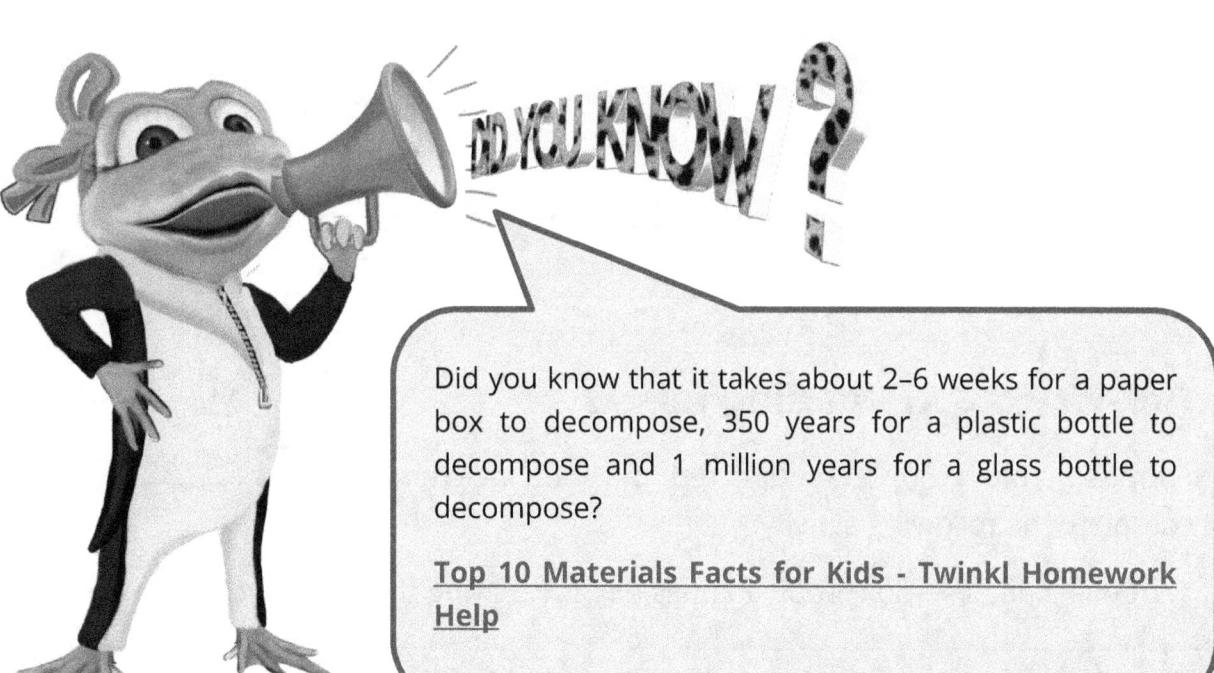

Did you know that it takes about 2-6 weeks for a paper box to decompose, 350 years for a plastic bottle to decompose and 1 million years for a glass bottle to decompose?

Top 10 Materials Facts for Kids - Twinkl Homework Help

Activity 6.3.1: Labels protect health

ATTENTION — Adult supervision required for this activity.

Working with a partner, complete the following activities:

Part 1

With your partner, research the effects of improper disposal of materials, such as electronic devices, plastic bottles and batteries. Prepare a poster to show how these materials harm the environment and human health. Your poster can be done digitally or manually.

Part 2

As Safe As Prepared (ASAP) is As Safe As Possible (ASAP) week at school

Using the slogan above as part of a safety awareness week at your school, create a safety audit of hazard points at school or home:

- staircase
- rails
- steps
- wet floors
- electrical outlets
- electrical panels
- appliances
- waste-water disposal
- exposed sharp objects and edges
- heavy objects likely to fall
- open trenches
- parking and assembly areas
- any others

Make recommendations to address or fix the situation.

Activity 6.3.2: Reading labels

ATTENTION — Adult supervision required for this activity.

Form suitable groups, as directed by your teacher, to complete the following activity.

1. Collect at least four product labels each and bring them to school.

Note: Labels with this symbol should not be taken to class. You can take a photograph of these labels to bring to class instead.

2. Within your group, carefully read each label. Record the following information for each of the different products in your notebook:

- instructions for use
- warning signs (hazard symbols)
- expiration date
- batch number
- ingredients
- storage
- disposal instructions.

3. List the similarities and differences you find between the labels.

4. Discuss the importance of the information shown on the labels, especially the information regarding the use, storage and disposal of each product.

Activity 6.3.3: Information and communication technologies (ICT) and me: Building a cool house

You are working on a construction site and are asked to research ways of keeping a building cool without using air conditioning. Research types of materials that are used to keep the temperature inside a building lower than outside the building.

Write a recommendation to the manager of your construction site telling him what you have found. Share your recommendation with your classmates to explain your findings.

Activity 6.3.4: Science, technology, engineering, arts and mathematics (STEAM) and me: Building a house

In groups of THREE, plan the construction of a miniature house from recycled materials using the engineering method. Think carefully about which materials would be best for each part of the house. Brainstorm and then decide what the house will look like, a step-by-step building procedure, the recycled material required and why you chose to use them.

Build your house together and present it, alongside your plan, to your class, explaining why you chose the materials you did. Were there any materials you would change if you were to do this project again? Why?

Activity 6.3.5 Crossword on properties of materials

Perform this activity in a group of four people.

Across

2. how hard it is to tear, bend or break something
5. the ability of a material to transfer heat

Down

1. a material we can see through
3. how well a material soaks up liquid
4. the ability of a material to attract another material

Evaluate yourself!

Use this evaluation grid to check your understanding of the concepts discussed in this chapter. Read each statement below and insert the symbol that best shows how well you feel you understand the concept. Ask a teacher or parent to help you go over any areas that are still unclear, or that you do not feel you have mastered. Be honest!

	I got it!	I need to do more work.	I do not get it. I need help.
I got it!	I need to do more work.	I do not get it. I need help.	

In this chapter:

	I got it!	I need to do more work.	I do not get it. I need help.
1. I can look at materials/objects in everyday use and tell if they are transparent, absorbent, have strength, have magnetic properties or conduct heat.			
2. I can list some properties of materials that determine the choice of objects for a specific purpose in everyday life.			
3. I can identify correct and safe ways of using, storing and disposing of materials and household items.			
4. I can evaluate how the disposal of selected materials affects the environment.			
5. I can assess the impact of different materials on society.			
6. I can generate predictions of material properties based on observations and experience.			

CHEETAH® PREPARING THE JAMAICAN SCIENTIST

How important is education to you? According to Mahatma Gandhi, 'True education is about getting the best out of oneself.'

Let's go, let's go, let's go! Let's leap to the next chapter. Let's leap for knowledge!

Chapter 7:
Reversible and irreversible changes

Chapter objectives

- ✓ Classify some changes as reversible and others as irreversible.
- ✓ Investigate to illustrate that some changes result in the formation of new materials and others do not.
- ✓ Infer that some materials can change from one state to another (solid, liquid and gas).
- ✓ Identify the processes involved when materials change from one state to another (freezing, melting, evaporating and condensing).
- ✓ Make careful observations of reversible and irreversible changes, record and explain these using suitable scientific language.
- ✓ Predict the effect of heat on selected materials.
- ✓ Predict whether a change will be reversible or irreversible.
- ✓ Test predictions of changes with actual observations.
- ✓ Distinguish between reversible and irreversible changes.

CHEETAH Science Fiction

Irreversible Change

Professor Jarvis had been a famous scientist for many years, working in the fields of science and mathematics. He was part of a team working on materials returned by the spaceship Mars I. One of these materials was a strange jelly-like liquid called 'Martian water.' He kept a small sample in his lab on the top shelf of his cupboard.

He did not realise that a broken pipe in his laboratory was leaking water onto the floor. He slipped on the wet floor, sending him flat on his back, hitting the shelf as he fell. This caused the bottle of Martian water to topple, spilling the jelly-like liquid onto the floor. Mr Jarvis was terrified, 'Oh my gosh!'

You see, in that instant, Mr Jarvis remembered the astronauts' warnings from a video interview. The Martian water samples came with a scary disclaimer: 'inert but unpredictable when mixed with Earth water'. Experiments by the world's leading scientists proved that when Martian water was mixed with Earth water, all seemed normal — at first. Then, an explosion occurred, giving off thick, brown fumes, which covered everything in its path, causing everything the fumes touched to decompose. When the vapour disappeared, everything that the vapour touched had been broken down into a black ash.

Spurred into action as he remembered what the astronauts said about these experiments, Mr Jarvis quickly grabbed a newspaper from the desk, trying to clean up the Earth water. He could not allow the water from the pipe to mix with the Martian water.

> **What safety protocols were not observed in Mr Jarvis's laboratory?**

CHEETAH® PREPARING THE JAMAICAN SCIENTIST

7.1: What are reversible and irreversible changes?

Change is a part of our daily lives. **Change** takes place when something has been made different from what it was before. We have four seasons of change in each year—spring, summer, autumn (fall) and winter; even the landscape changes over thousands of years. We change from babies into adolescents, then into adults over time.

Some changes can be temporary, while other changes can be permanent. These changes are called **reversible** and **irreversible changes.**

Reversible changes

A reversible change is a change that *can* be undone. That means the thing that was changed can go back to its original substance, or to what it was before. Reversible changes can be temporary. When a reversible change is made, we can go back to the starting material. A reversible change might alter how the material looks and feels, but it does not make any new material. Examples of reversible changes are **physical changes** such as the bending of a wire or the crushing of paper. Dissolving, evaporating, melting and freezing are all examples of reversible changes.

Let us review one example of a reversible physical change:

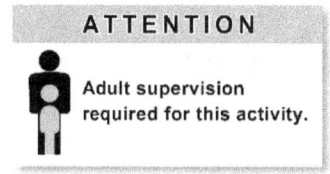

You can (**1**) —dissolve salt crystals in water, then (**2**) —heat the mixture to boiling so that most of the water evaporates, (**3**) — remove the mixture from the heat and pour it into an evaporating dish to allow the mixture to air dry as the remaining water evaporates, then last (**4**) — place the salt reformed into a clean dry container. This is a reversible change.

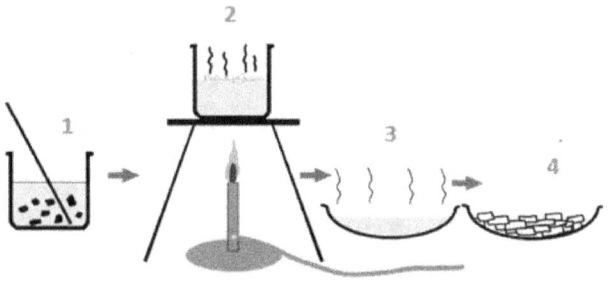

Figure 7.1: Reversible change.

Reversible changes and states of matter

You know that matter exists in three states: solids, liquids and gases. When matter changes from one state to another, it is undergoing a reversible change. The processes that change a material's state of matter include **freezing**, **melting**, **evaporating** and **condensing**.

- **Freezing** happens when liquid changes into a solid by losing heat. For example, when liquid water is put into a freezer, it becomes ice.

- **Melting** happens when solid changes into a liquid by gaining heat. It is the opposite of freezing. This is why ice will turn back into liquid water after it is taken out of the freezer.

- **Evaporation** happens when liquid changes into a gas by gaining heat. When water is boiled or heated, it creates steam, a vapour or gaseous form of liquid water, which looks like a white mist.

- **Condensation** takes place when gas changes into liquid by losing heat. Water vapour cools and forms tiny droplets of water in the air. This is why droplets of water will form on the outside of a cold glass when it is warm.

State changes in water

Many substances can be melted and frozen. All substances have a melting point and a boiling point. Therefore, all substances can exist in three states, depending on the temperature.

Water, for example, exists in three states: solid (ice), liquid (water) and gas or vapour (steam). The conversion of water between the three states involves reversible changes.

- When ice gains heat, it melts to form water.
- When water gains heat, it boils and evaporates to form water vapour.
- Water vapour can then lose heat and condense to become water again.
- Water can also lose heat and freeze to re-form into ice.

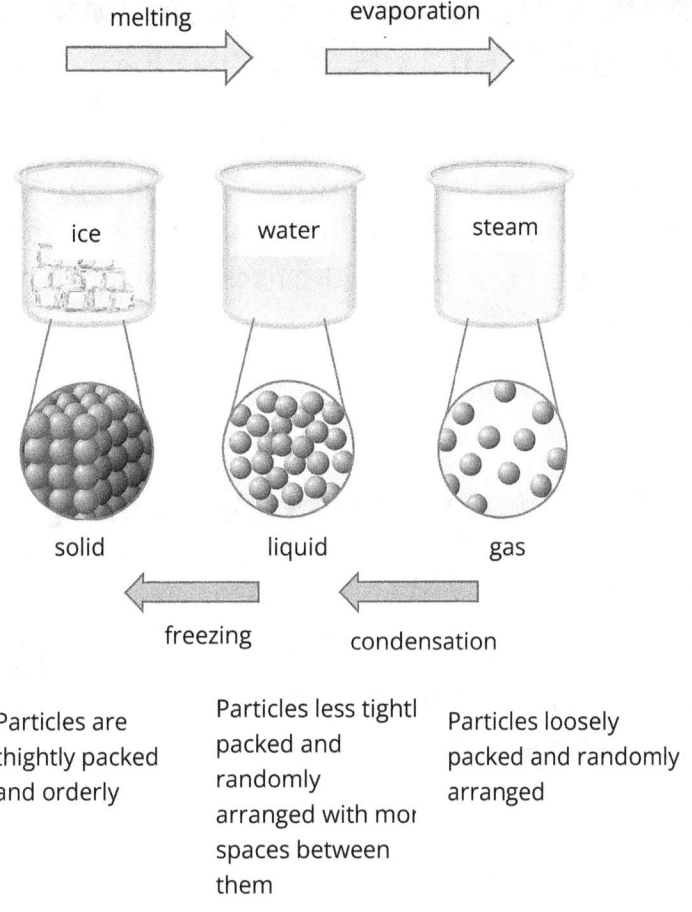

Figure 7.2: Changes in the state of water.

Some examples of reversible changes are:

- bending a straw
- dissolving salt in water
- stretching an elastic band
- inflating a balloon
- mixing oil and water
- melting ice
- freezing water
- condensation of water
- boiling of water
- evaporation of water.

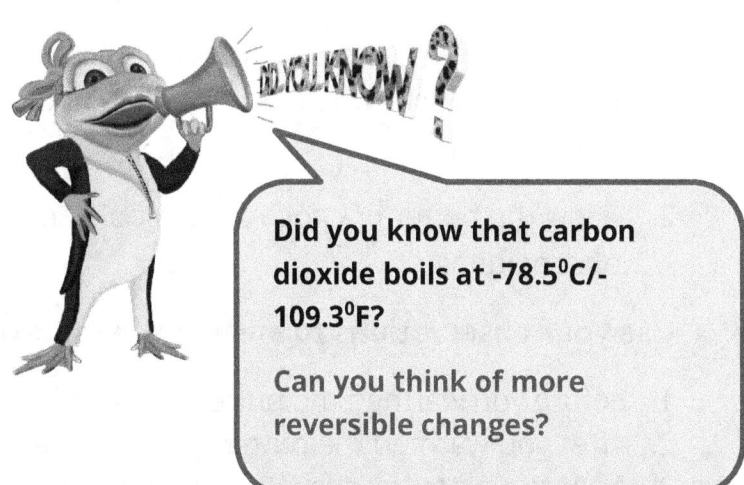

Did you know that carbon dioxide boils at $-78.5^0C/-109.3^0F$?

Can you think of more reversible changes?

Activity 7.1.1: Information and communication technologies (ICT) and me: Before and after a reversible change

As you complete Activity 7.2 with a partner, take turns taking digital photos of the items you test. Show the starting products and the final products. Together, use these images to create a presentation discussing the meaning of reversible change. Share your results with your classmates.

Activity 7.1.2: Exploring reversible changes

Work with a partner to perform the following experiments:

SAFETY NOTE: Students should not use hot water without careful adult supervision.

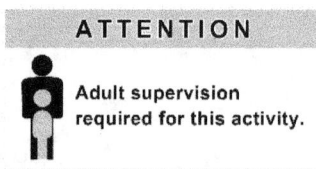

Part 1: Melting

Materials:

- butter or chocolate
- beaker
- bowl
- hot water
- butter knife

Procedure:

1. Gently heat a cube of butter or three pieces of chocolate in a container over another container of hot water, as shown in the diagram. Record your observations.

2. Allow the butter or chocolate to cool. Record your observations.

Use your observations to answer the following questions:

1. Before you warmed the butter or chocolate, was it a solid, liquid or gas?
2. After you warmed the butter or chocolate, was it a solid, liquid or gas?
3. After the butter or chocolate cooled, was it a solid, liquid or gas?

Part 2a: Evaporation and condensation

Materials:

- kettle
- container
- pot cover or glass
- water
- kitchen mitts (gloves)
- heat source

Procedure:

1. Put water in the kettle and bring it to a boil over a slow fire.
2. Place a pot cover directly over the steam from the kettle of water.
3. Allow the water formed to run into another container.
4. Record your observations.

Use your observations to answer the following questions:

1. Describe the change that occurred inside the pot.
2. Describe the change that took place on the surface of the pot cover facing the steam.
3. Why did the steam change to water?
4. Why was it important for the water to boil?
5. Why is the water inside the container pure water?

Part 3: Condensation on a cold glass

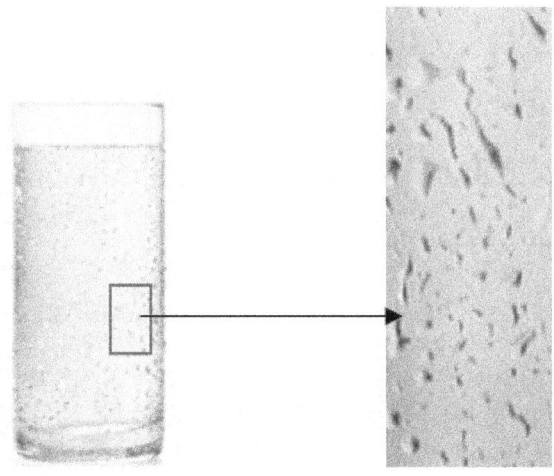

Materials:

- glass
- cold water

Procedure:

1. Pour cold water into a glass and leave it to sit at room temperature (25–30^{0}C)
2. Record your observations.

Use your observations to answer the following questions:

1. What changes do you see on the outer surface of the glass?
2. Where did the water come from?
3. What is the process that caused the changes that you saw?
4. Why did it take place on the outside of the glass?
5. Suppose hot water was placed in the glass, would you have seen the same changes?
6. Would the same observations be made if this experiment was done in Antarctica? Explain why or why not.
7. Why was the windscreen of the car fogged up on the inside of the car and not the outside?

Part 4: Folding of paper

Materials:

- letter-size paper
- newspaper (one sheet)

Procedure:

1. Fold a letter-size paper or a sheet of newspaper into four parts. Then open it.
2. Record your observations.

Use your observations to answer the following questions:

1. What changes do you see in the paper?
2. Did the size of the paper change after it was folded and unfolded?
3. Did the amount of material you used change in any part of your investigation? Did any new materials form?

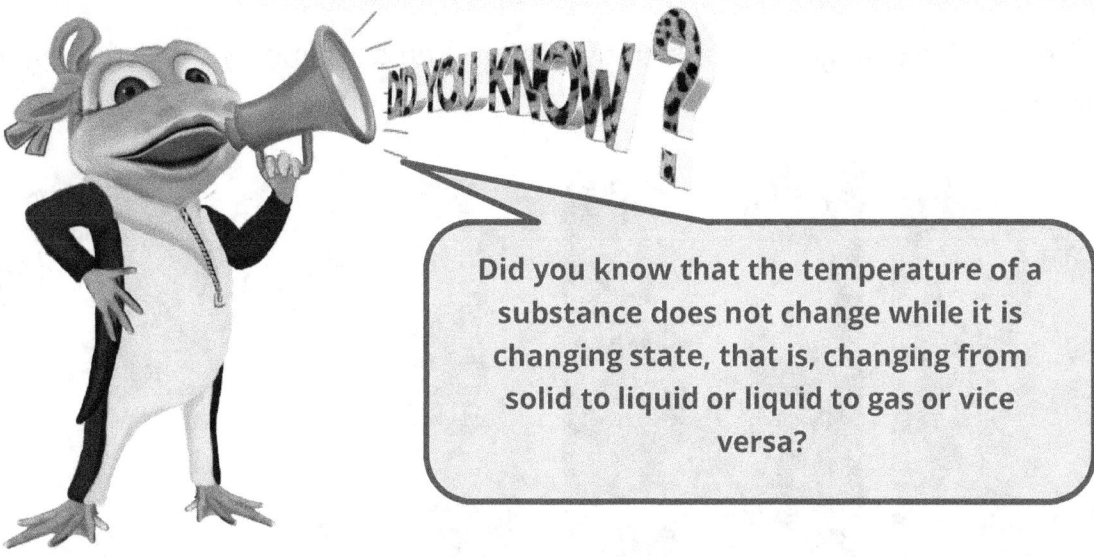

Did you know that the temperature of a substance does not change while it is changing state, that is, changing from solid to liquid or liquid to gas or vice versa?

Irreversible changes

Growth is an irreversible change because you cannot become a baby again. Remember, an irreversible change means things cannot go back to the starting point. Irreversible changes are permanent changes.

Figure 7.3: *Example of irreversible change (growth).*

Other irreversible changes are caused by mixing chemicals and by the action of heat on mixtures.

Figure 7.4: *Examples of irreversible changes: (from left) ice cream, fire burning, mixing ingredients to make bread.*

For example, when oil, yeast, sugar, flour and water are combined and heated, the heated mixture forms bread. Bread cannot be separated to give oil, yeast, sugar, flour and water again. Bread, therefore, is a new substance that is formed and so it is a permanent change.

Whenever a new substance is formed, we have an irreversible change. For example, if you burn paper, you get ash. Ash is different from the paper you started with. There is nothing that you can do to recover the paper once it is set on fire.

Some other examples of irreversible changes are:

- burning petrol in vehicles
- rusting iron
- cooking food
- burning wood
- mixing vinegar and baking soda
- exposing a peeled green banana to air
- rotting (decomposition).

Activity 7.1.3: Information and communication technologies (ICT) and me: Before and after an irreversible change

As you complete Activity 7.4 with a partner, take turns taking digital photos of the items you tested. Show the starting products and the final products. Together, use these images to create a presentation discussing the meaning of irreversible change. Share your results with your classmates.

Activity 7.1.4   Exploring irreversible changes

Work with a partner to perform the following experiments:

Part 2: Burning paper

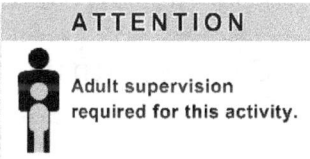

Safety note: Students should not use matches or flames without careful adult supervision.

Materials:

- a small piece of newspaper
- metal container
- matches.

Procedure:

1. Burn a small piece of newspaper in a metal container outdoors.
2. Record your observations before, during and after the burning is completed.

Questions:

1. What kind of change have you observed?
2. Do you think we could collect the smoke and the ash and change this back into a newspaper?
3. What happened when you burned the pieces of newspaper in the metal container?

Paper burning in a metal bin

CHEETAH® PREPARING THE JAMAICAN SCIENTIST

Part 3: Mixing vinegar and baking soda

Test A

Materials:

- baking soda
- vinegar
- one large balloon
- 500/600 ml plastic bottle

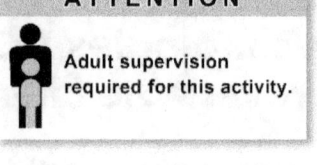

ATTENTION

Adult supervision required for this activity.

Procedure:

1. Place two tablespoons of baking soda into a balloon.

2. Pour one-third of a cup of vinegar into a plastic bottle.

3. Gently stretch the balloon over the mouth of the bottle. Make sure that none of the baking soda in the balloon spills into the plastic bottle as you are stretching it over the top.

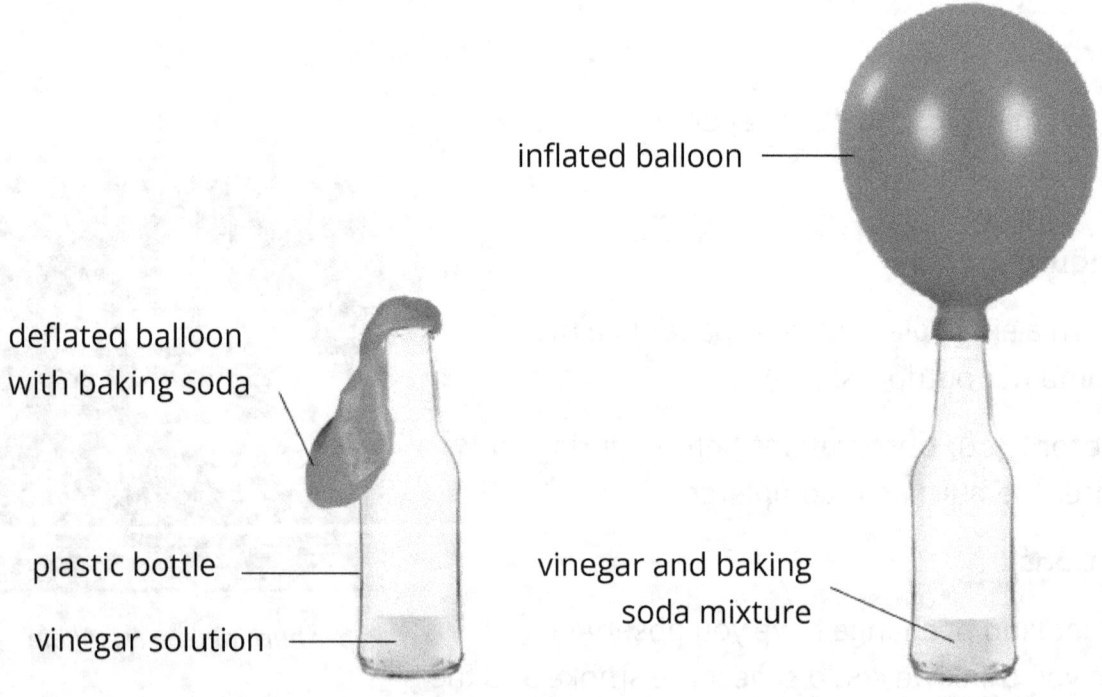

4. Hold the tip of the balloon and gently lift it above the mouth of the bottle. Allow the baking soda to fall into the bottle. Record your observations.

Use your observations to answer the following questions:

1. What caused the balloon to become inflated?
2. What new material(s) was/were formed in this experiment?
3. What would happen if you increased the amounts of vinegar and baking soda used for this experiment?
4. If we had put the vinegar in the balloon, would we have gotten a different result?
5. Research the name of the gas formed when baking soda and vinegar are mixed.

Let's review reversible and irreversible changes:

1. What is a reversible change?
2. What is an irreversible change?
3. Give three examples of each type of change.
4. Why is cutting up a bit of paper a reversible change?
5. Your brother left a bowl of milk on the kitchen counter overnight. In the morning, he observed a thick lumpy curd settling on top. Your brother is sad and wonders what he could do to bring back the milk. Explain to your brother why this is not a reversible change.
6. Complete the diagram below by inserting the name of the processes at A, B, C and D.

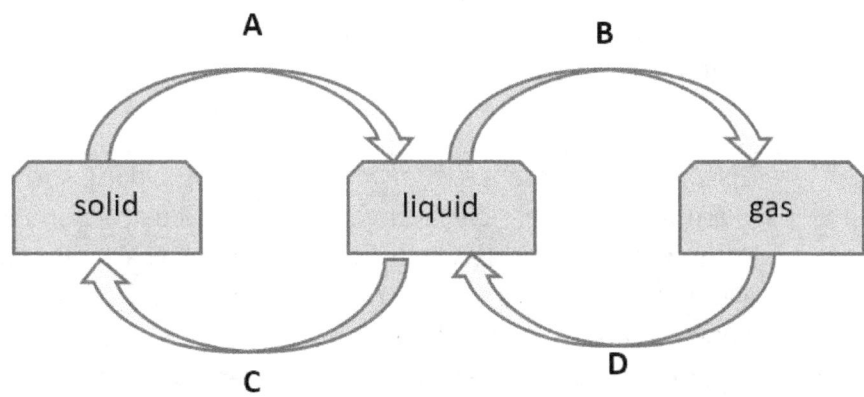

7.2: Comparing changes

Comparing reversible and irreversible changes

The table below lists the major differences between reversible and irreversible changes.

Different features	Reversible change	Irreversible change
Permanence	• Temporary change. The material can return to its original form.	• Permanent change. The material cannot return to its original state.
End product	• The end product is often similar to the original material but can be different too. The end product is usually a blend of materials. For example, sugar and water look like water depending on the colour of the sugar used. The end product can also be from a change of state, as in the case of melted butter or chocolate.	• The new product is different from the original material. These changes are sometimes accompanied by a change of colour. For example, when steel rusts, it makes the irreversible change of becoming brown and crumbly. Sometimes gas is given off, such as when vinegar is mixed with baking soda.
Adding and removing heat	• Melting is a reversible change. Melted butter will change back into solid butter by cooling. • Freezing is an example of a reversible change. If you freeze water to make ice, the ice can be melted to make water.	• Heat can cause an irreversible change. A boiled egg cannot be changed back to a raw egg. • Burning is an example of an irreversible change. Whenever paper is burnt, new materials such as ash and smoke are produced.
Making mixtures	• Dissolving a substance is an example of a reversible change. When sugar or salt dissolves in water, it creates sugar or salt water. It can be changed back to sugar or salt and water using evaporation.	• Mixing baking soda with vinegar produces carbon dioxide. The mixture cannot return to vinegar and baking soda because bubbles of carbon dioxide gas were given off.

CHEETAH™ collaboration corner

Activity 7.2.1: Science, technology, engineering, arts and mathematics (STEAM) and me: Making plastics

Form a group of three to investigate how plastics form.

Materials:

- 200 ml of milk
- 5% vinegar solution
- paper towel
- small pot
- strainer
- olive oil

Procedure:

1. Pour 200 ml of milk into a small pot, then heat the milk for four minutes.
2. Add a teaspoon of vinegar to the heated milk. Record your observations.
3. Separate the solid curds from the milk using a strainer and take turns placing them on a paper towel or a plain piece of paper. Observe and discuss their texture as you do this. Let them remain there for ten minutes.
4. Mould the `plastic' into a shape of your choice, ensuring every person gets the opportunity to feel how the texture has changed. Place the moulded shape in an open area and allow it to dry for twelve hours.
5. Add a thin layer of olive oil to the surface of the mould.
6. Each person should feel the moulded shape once more. Discuss and record your observations.

Use your observations to answer the following questions:

1. Is it possible to convert the solid curds back to vinegar and milk? Explain your answer.
2. Did this experiment demonstrate a reversible or irreversible change? Explain your answer.
3. How did the materials in this experiment change from one state to another?

Activity 7.2.2: Information and communication technologies (ICT) and me: Changes around me

The rusting of iron wastes materials and money. Research and create a presentation on the uses of iron in your household and community. State how the rusting of iron affects the durability of products and possible ways to stop or slow down rusting.

DID YOU KNOW?

Did you know that the lava from a volcano can be between 1300 to 2200 ^{0}F?

Activity 7.2.3: Crossword on reversible and irreversible changes

Using the clues below, completed the following crossword puzzle.

Across
- 3. change from liquid to solid
- 5. matter that makes a solid
- 8. can go back to the original
- 10. vibrations moving through air or water that can be heard
- 12. solid, liquid or gas (3 words)

Down
- 1. change from gas to liquid
- 2. cannot go back to the original
- 4. boiling water changes to it
- 6. firm, stable shape
- 7. melted solid
- 9. change from solid to liquid
- 11. change from liquid to gas

CHEETAH® PREPARING THE JAMAICAN SCIENTIST

Evaluate yourself!

Use this evaluation grid to check your understanding of the concepts discussed in this chapter. Read each statement below and insert the symbol that best shows how well you feel you understand the concept. Ask a teacher or parent to help you go over any areas that are still unclear, or that you do not feel you have mastered. Be honest!

I got it!   I need to do more work.   I do not get it. I need help.

In this chapter:

	I got it!	I need to do more work.	I do not get it. I need help.
1. I can group some changes as reversible and others as irreversible.			
2. I can do an investigation to show that some changes cause the formation of new materials and others do not.			
3. I understand that some materials can change from one state to another (solid, liquid and gas).			
4. I can identify the processes involved when materials change from one state to another (freezing, melting, evaporating and condensing).			
5. I can make careful observations of reversible and irreversible changes, record			

	I got it!	I need to do more work.	I do not get it. I need help.
and explain these using suitable scientific language.			
6. I can predict the effect of heat on selected materials.			
7. I can predict whether a change will be reversible or irreversible.			
8. I can test predictions of changes with actual observations.			
9. I can distinguish between reversible and irreversible changes.			

Grab your Merge Cube to explore the augmented reality activities next.

Augmented reality: Structure of matter

Matter is anything that has mass and takes up space. In these activities, you'll explore what makes one kind of matter different from another. For example, what makes sand different from rubber? Then you will use what you have learned to build some neat structures. Let us begin!

Activity 1

Material creator: All solid, liquid and gaseous substances are matter, but not all matter is the same. Create all kinds of materials in this activity by changing properties like hardness, weight and surface type.

Activity 2

Block craft: The properties of some types of matter make them a better fit for certain uses, like construction. Now try out the materials you made in the previous activity to see how they stack up.

According to Christian Lous Lange, 'Technology is a useful servant but a dangerous master.' What do you think this means?

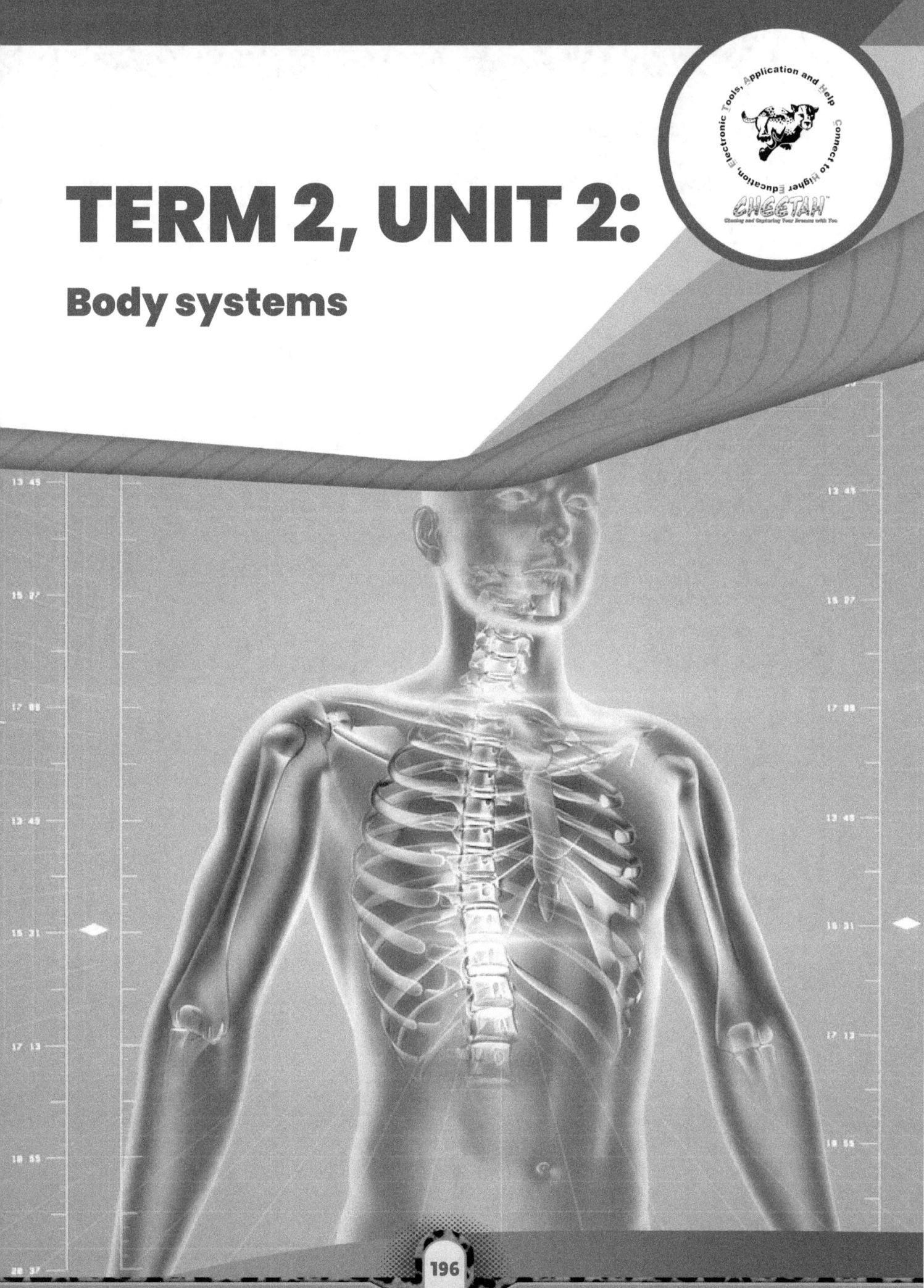

TERM 2, UNIT 2:
Body systems

Words worth knowing

The list below contains the scientific words that will be discussed in this unit.

- arteries
- blood
- blood vessels
- bone marrow
- bones
- brain
- circulatory system
- colon
- diaphragm
- digestion
- egg
- excretion
- excretory system
- exhalation
- fertilisation
- heart
- homeostasis
- inhalation
- joints
- kidney
- large intestine
- larynx
- ligaments
- lungs
- mouth
- muscles
- musculoskeletal system
- nerves
- nervous system
- nose
- oesophagus
- organ
- organ system
- reproductive system
- respiration
- respiratory system
- skin
- small intestine
- sperm
- spinal cord
- stomach
- teeth
- tendons
- tissues
- tongue
- trachea
- uterus
- veins

Chapter 8: Human body system

Chapter objectives

Specific objectives:
- Explain what is meant by the term system.
- Identify the organ systems and state their functions in humans.
- Recognise the integration of the different organ systems in carrying out life processes.
- Identify selected organs in the human digestive system (mouth, oesophagus, stomach, small intestine and large intestine) and outline the path food travels from mouth to anus.
- Describe how the bones, joints and muscles work together to produce movement in humans.
- Identify the excretory organs of humans (kidneys, lungs and skin) and state their role in excretion.

CHEETAH Science Fiction

Journey Inside the Body

Croaky Croak, the frog, sat on a log, rubbing her head. She had crashed into a tree and fallen off her bike.

'You didn't wear a cycle helmet,' said LaChase, the cheetah. 'You should look after your body at all times. And not just the outside. There are lots of things happening inside. Our bodies are constantly working to keep us alive. We should do all we can to protect our bodies.'

Croaky looked confused.

'I'll explain,' said LaChase. 'Let's play a game. Imagine we are in a tiny submarine attached to an oxygen molecule in the air. Now let's go for a ride inside the human body! We will travel through the nostrils and enter through the bloodstream into the lungs.'

'Perfect,' said Croaky. Croaky closed her eyes. When she opened them, she was looking out of a window. 'Everything is red out there!' she said.

'That's nothing to be worried about,' said LaChase. 'We're in the bloodstream. Let's go to the brain and look around. Look at all those loops and strings. They are very delicate. A helmet would help protect these fragile nerves if there is a crash.'

Going up and down steep slopes, they swished past joints and bones. Soon, they could hear the stomach churning and feel the rhythmic vibrations of the heart as they went past the chest. As they looked in the lower back, they could see the filter pumps of the kidneys working. They did not want to be filtered so LaChase quickly turned the steering wheel and sent them going down the leg. They stopped briefly in the tissues and looked around on the toes. It was hard work to come back up the leg, where they saw large muscles attached to long bones; but the blood pushed them back up.

'I never thought our bodies were so amazing,' said Croaky.

'Even when we're sleeping, our body systems continue to work,' said LaChase.

'Wow! What a journey, but how do we get out of here?'

'We have to attach our submarine to carbon dioxide to reach the lungs so we can go back home.'

They closed their eyes and when they opened them, they were both sitting on the log. Croaky rubbed her body all over, imagining everything that was happening inside her.

'From now on,' she said, 'I'm going to start caring for my body a lot more.'

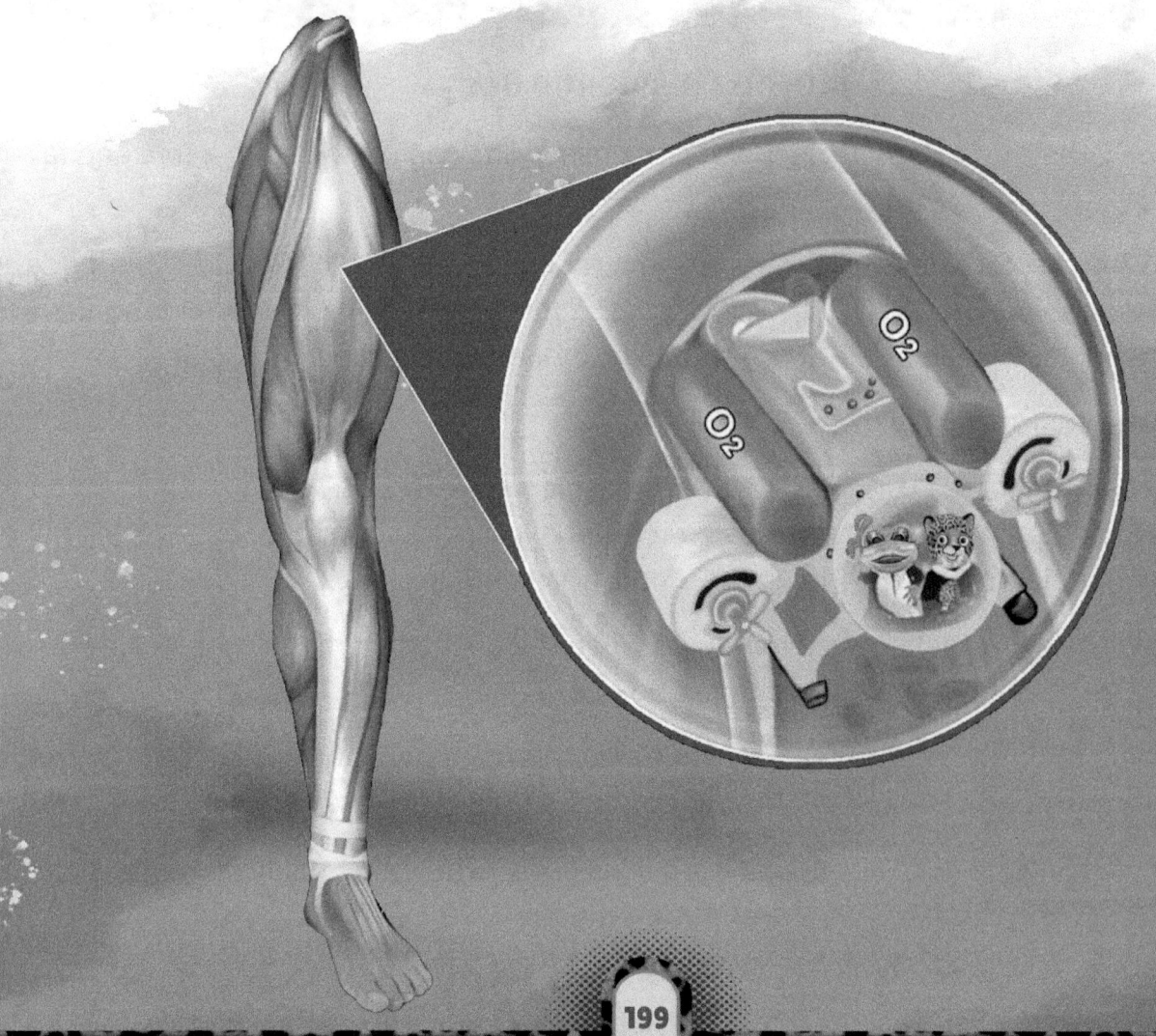

8.1: What is a body system?

Organ systems

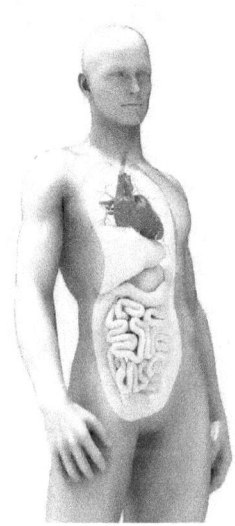

Figure 8.1: Organ systems.

Have you ever wondered how a car works? You can see the outside but do you know that numerous parts are working together on the inside to make it move? Like the car, a body has systems. A system is a set of parts that work together to perform a specific task.

The body is like a finely tuned machine. It is made of many organs that work together to carry out special functions. These organs work through interconnected networks as a system called an organ system. It takes all the organ

Figure 8.2: kids playing sport.

systems in the body to work together to keep us alive and well. Let us look closely at how our body works.

An organ is a body structure that has a specific function. Some of the organs are the heart, stomach, skin and kidneys. The heart pumps blood around the body. The stomach helps us to digest food. The skin excretes wastes and protects the body from infection, and the kidneys filter off waste materials that are excreted in the urine.

The stomach, pancreas, intestine and colon work together to form the digestive organ system. These organs work together to carry out the specific function of breaking down the food we eat so the body can absorb and use it.

The body's organ systems are the:

- digestive system
- musculoskeletal system
- excretory system
- circulatory system
- respiratory system
- reproductive system
- nervous system.

Table 8.1: Functions of the human body's organ systems

Organ systems	Interdependent organs within each system	Function
Digestive system	Mouth, stomach, pancreas, small and large intestine	Helps to break down food in a form that the body finds useful.
Musculoskeletal system	Bones, joints and muscles	Works together to allow the body to move, and also protects major organs within the body.
Excretory system	Kidney, bladder, lungs and skin	Removes substances that may be harmful to the body.
Circulatory system	Heart, blood and blood vessels	Pumps blood through the body and transports materials such as oxygen, carbon dioxide, nutrients and water to different parts of the body and removes wastes.
Respiratory system	Nose, trachea and lungs	Exchanges gases such as oxygen and carbon dioxide between the body's cells. Gets/Provides energy from the food we eat.
Reproductive system	Penis, testes, vagina, ovary and uterus	Produces offspring.
Nervous system	Brain, spinal cord and nerves	Controls the body's actions.

Integration of the different organ systems

Today is a big day for Paul. He is competing in champs at the National Stadium. He had a nutritious breakfast and had trained, but that did not stop him from being nervous. He started to sweat and was feeling a little nervous. He heard the whistle and he was off like lightning. Which organ systems does he need?

To show that systems work together to achieve a single goal, considering the way Paul gets energy from the food he eats is a good example of systems working together to achieve one goal (Paul running fast and winning the race). He needs help from at least six organ systems working together: the digestive system, the respiratory system (lungs), the nervous system, the endocrine system, the circulatory system and the musculoskeletal system.

The digestive system produces glucose (nutrients) from breakfast. As he breathes, the respiratory system provides oxygen. The nervous system allows him to hear the whistle and move his joints quickly. The circulatory system carries both glucose and oxygen to the muscle cells, where respiration takes place. All systems work together and allow him to win the race.

This chapter will explore the digestive system, the musculoskeletal system and the excretory system.

Let's explore these systems together!

8.2: The human digestive system

What is digestion?

Digestion occurs when the food we eat is broken down to give our body the nutrients it needs. Digested food gives the body the protein, fats, carbohydrates, vitamins and minerals it needs to function properly and repair itself. The digestive process also provides roughage, which helps remove waste from the digestive system.

Parts of the digestive system

The digestive system is made up of the following components:

Mouth: Food is taken into the mouth by a process known as ingestion. Digestion begins in the mouth. When food enters the mouth, the **teeth** crush the food to break it down into smaller pieces. Special chemicals in the saliva begin breaking down food into its major nutrients. Once the food has been crushed, the **tongue** pushes the soft mush into the throat.

Oesophagus: After the food enters the throat, it is swallowed down toward the oesophagus. The oesophagus is a long, straight, muscular tube that contracts and relaxes to squeeze food into the stomach.

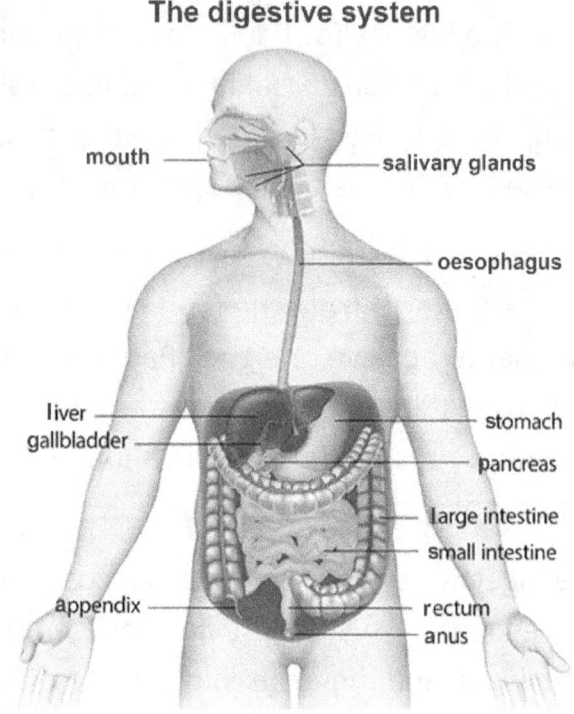

Figure 8.3: *The human digestive system.*

Stomach: The stomach is a muscular, bag-like organ located at the end of the oesophagus. The stomach squeezes the food and mixes it with acids. This helps to break down the food so the body can absorb it.

Small intestine: The food leaves the stomach and enters the small intestine, where digestion is completed. Even though it is called the small intestine, this organ is approximately six metres long. The small intestine absorbs most of the nutrients (glucose, amino acids, fatty acids, vitamins and minerals) that the body needs. All the remaining material is waste, which is passed to the large intestine.

Large intestine: The large intestine absorbs water from the waste, stores the waste in the rectum and passes it through the anus.

Did you know food takes about seven seconds to travel down the oesophagus?

Activity 8.2.1: Science, technology, engineering, arts and mathematics (STEAM) and me: The digestive system

Plan, design and construct a model of the digestive system using available materials from your home. Make sure to include the major organs. Create a flow diagram that shows how food travels through the digestive system.

Remember to use the engineering method to make the model.

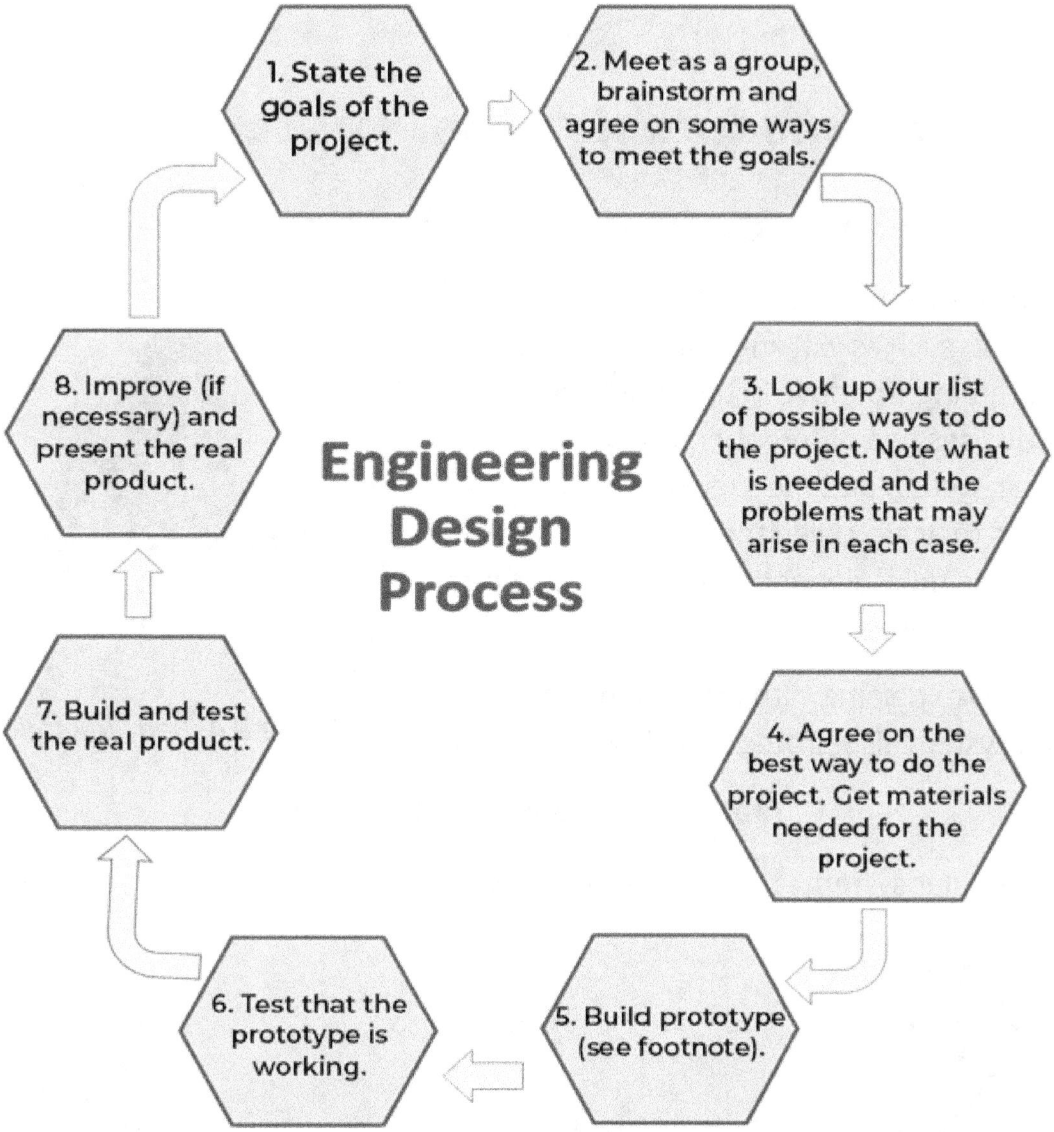

Engineering Design Process

1. State the goals of the project.
2. Meet as a group, brainstorm and agree on some ways to meet the goals.
3. Look up your list of possible ways to do the project. Note what is needed and the problems that may arise in each case.
4. Agree on the best way to do the project. Get materials needed for the project.
5. Build prototype (see footnote).
6. Test that the prototype is working.
7. Build and test the real product.
8. Improve (if necessary) and present the real product.

Note: Step 5. Prototypes or trial versions may be built and tested using relatively inexpensive materials to avoid wasting resources. For example, using paper instead of cardboard or hardwood or using glue instead of cement to reduce expense. Step 6. If the prototype tested does not work as it should, you may need to go back to the brainstorm step (#2).

Let's examine the digestive system.

1. What are the main organs of the digestive system?

2. Where does digestion begin?

3. Where does digestion end?

4. Why is water removed in the large intestine?

8.3: The musculoskeletal system

What is the musculoskeletal system?

The body performs many different types of movement, such as walking, running, jumping and swimming. To move in these different ways, the body uses two different organ systems: the muscular system and the skeletal system. Together, they are called the musculoskeletal system. The muscular and skeletal systems put the body in motion.

In addition to helping with movement, the skeletal system also protects the body's organs.

Figure 8.4: *Body in motion.*

Table 8.2: Musculoskeletal system

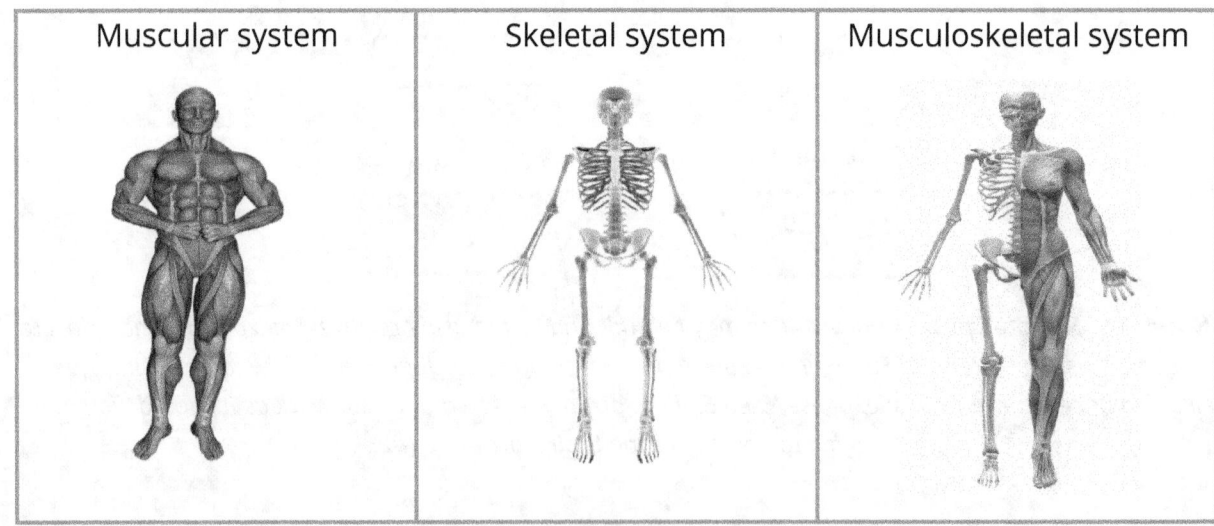

| Muscular system | Skeletal system | Musculoskeletal system |

Parts of the musculoskeletal system

Bones are the major organs of the skeletal system. They provide a structure for the muscles to attach to, as well as support and protect the body's other organs. Within the centre of bones, there is a soft tissue called **bone marrow**. Bone marrow helps to produce new blood cells and stores fat.

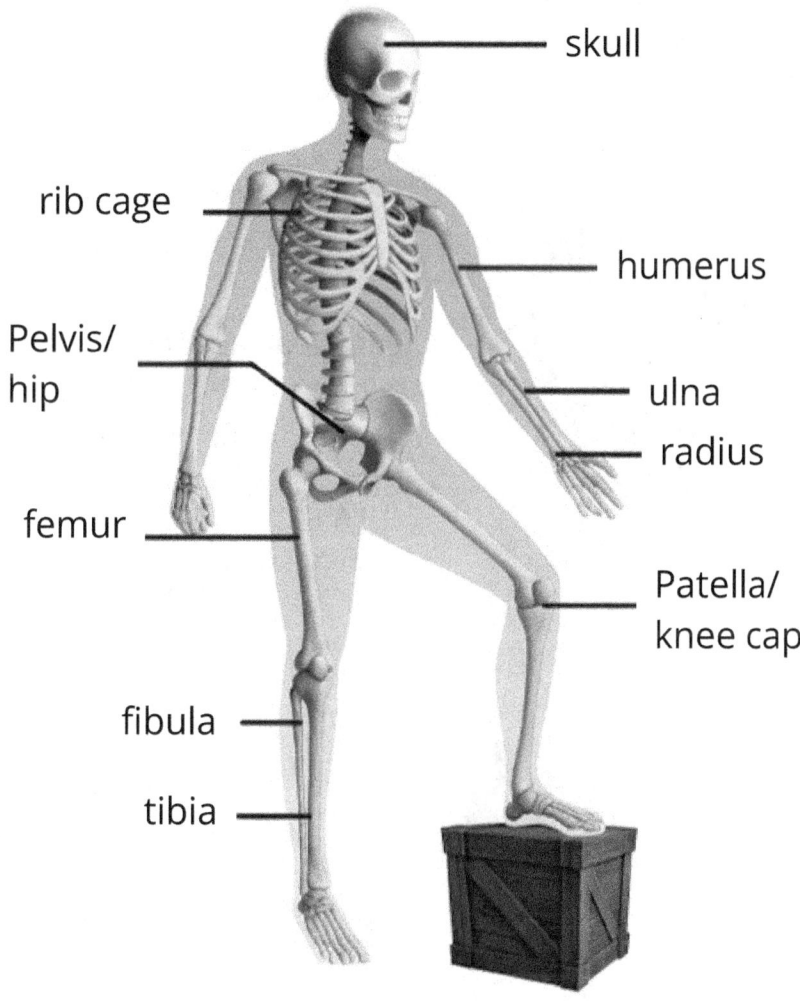

Figure 8.5: *The human skeletal system.*

Muscles are the main organs of the muscular system. They are controlled by the nervous system. Their primary function is to move the skeleton. Bones are fastened to muscles by strong cords of tissue called **tendons** (Figure 8.3).

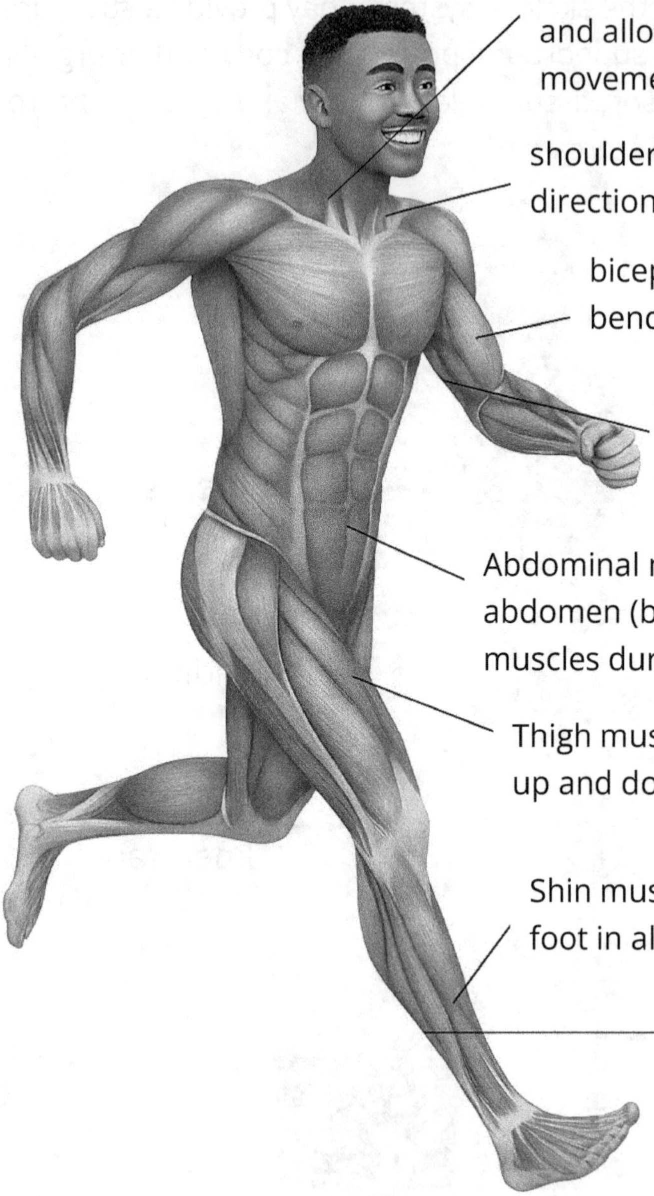

- neck muscles - hold head upright and allows up, down, left right movements
- shoulder muscles - move the arm in all directions
- bicep muscles – contract and bend the arm
- Tricep muscles – contract and straighten the arm
- Abdominal muscles- move the abdomen (belly) and help chest muscles during breathing
- Thigh muscles- move the leg up and down
- Shin muscles- help move the foot in all directions
- Calf muscles – flex the foot by moving the heel, pointing the toes and rotating the ankle

Figure 8.6: Muscles in the human body.

Muscles work by pulling against each other. For a joint to move in two directions, two muscles that pull in opposite directions are required. As one muscle contracts (shortens), the other relaxes (extends).

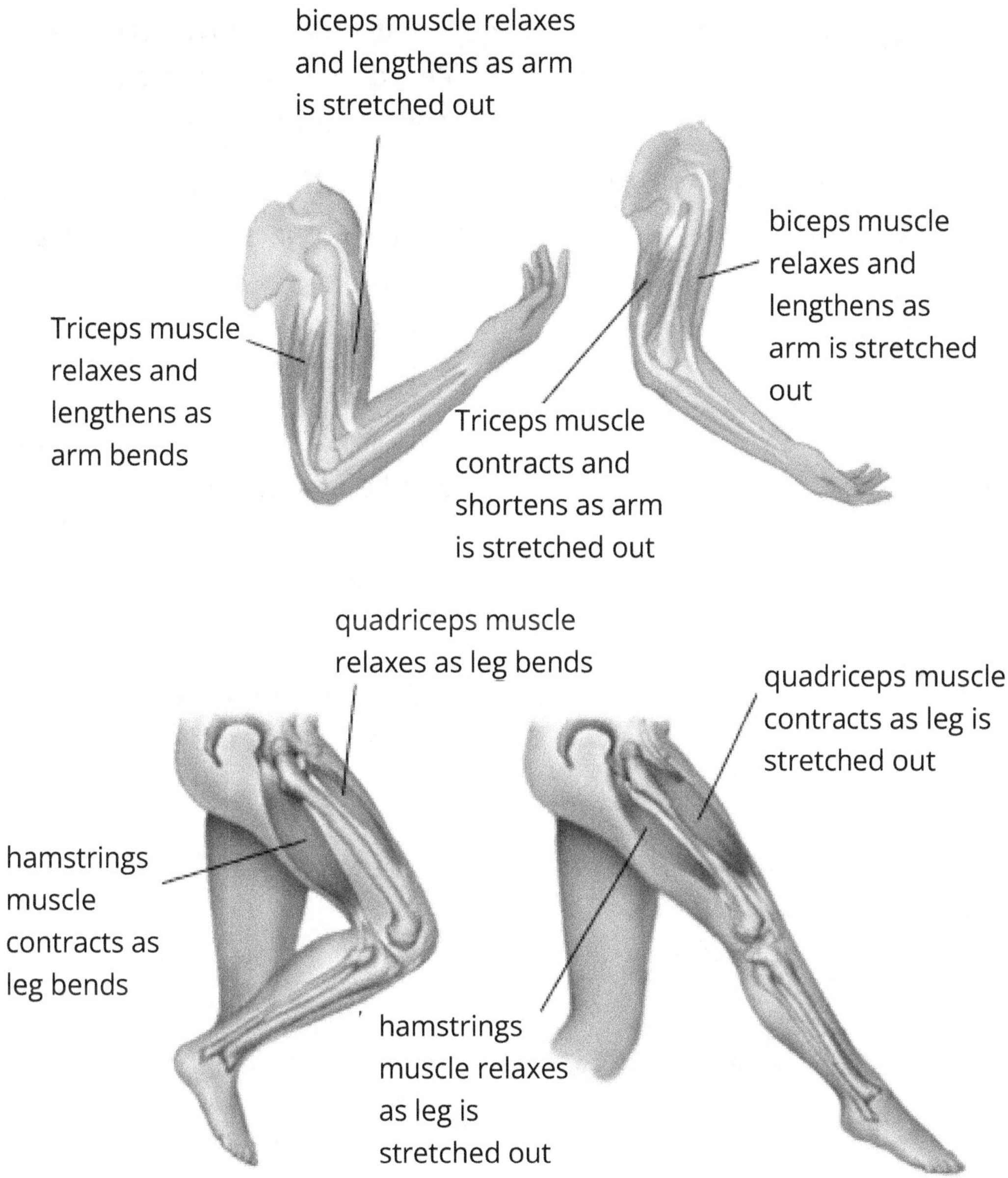

Figure 8.7: Muscles pulling against each other in the arm and leg.

Joints occur when two bones meet each other in the body. They help to hold the skeleton together and allow the body to move when the skeletal muscles contract. Joints are kept together with strong elastic tissues called **ligaments,** which hold the joints in place. Some joints do not move at all, such as the joint where a tooth connects to the jawbone. Some joints can move in multiple directions, such as the hip and shoulder.

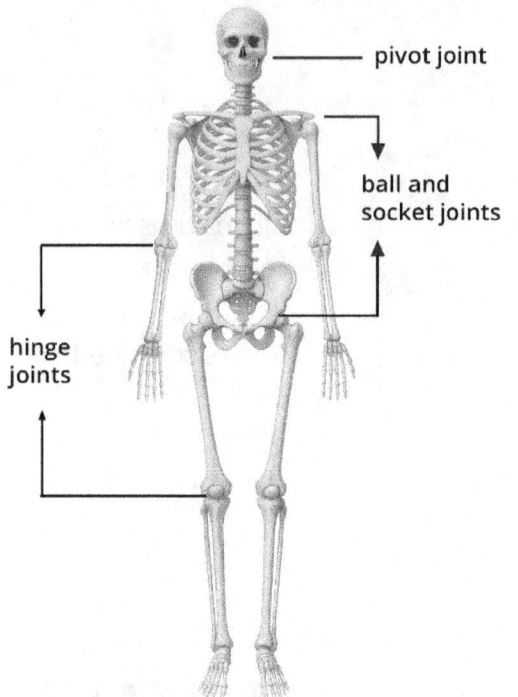

Figure 8.8: Joints in the human body.

- **Hinge joints** move in one direction, like a door opening and closing, e.g. your elbows and knees.

- **Pivot joints** move in a turning or twisting motion, like screwing and unscrewing the lid of a bottle. For example, there is a pivot joint in the neck.

- **Ball-and-socket joints** are where the round end of a bone fits into a socket. They allow movement in all directions. The shoulders and hips are examples.

Cartilage is a soft, gel-like padding found between two bones. It protects joints and helps movement.

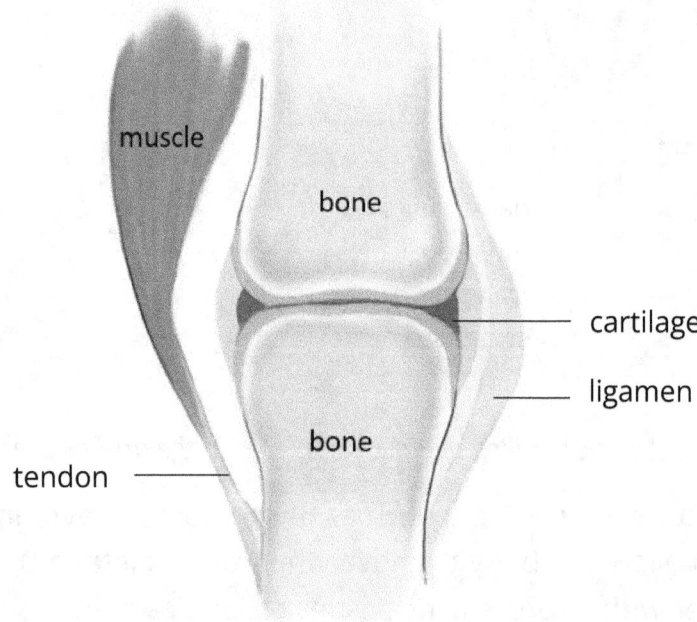

Figure 8.9: Cartilage, ligaments and tendons.

Activity 8.3.1: Exploring joints

With a partner, think about the joints in your body and explore how they move. Complete the table, putting a tick in each row to classify the joints as hinge, pivot or ball and socket.

Joint	Hinge	Pivot	Ball and socket
Knee			
Hip			
Neck			
Elbow			
Shoulder			

Let's examine the musculoskeletal system.

1. What are the main organs of the musculoskeletal system?
2. What do you think forms a cushion between two bones?
3. How do the biceps and the triceps work together to lift a box?
4. In a race, an athlete hurts the hamstring muscle. What part of his body does he hold in pain?
5. Research damage to the joints caused by arthritis.

Did you know that when you are asleep, your brain is sometimes more active than when you are awake?

8.4: The excretory system

Our bodies are always trying to remain in a state of balance. In order to do that, we must get rid of any harmful substances that build up in our bodies as we work and play. These are called *waste products*. **Excretion** is the removal of waste through the excretory system. The human excretory system includes the kidneys, lungs and skin.

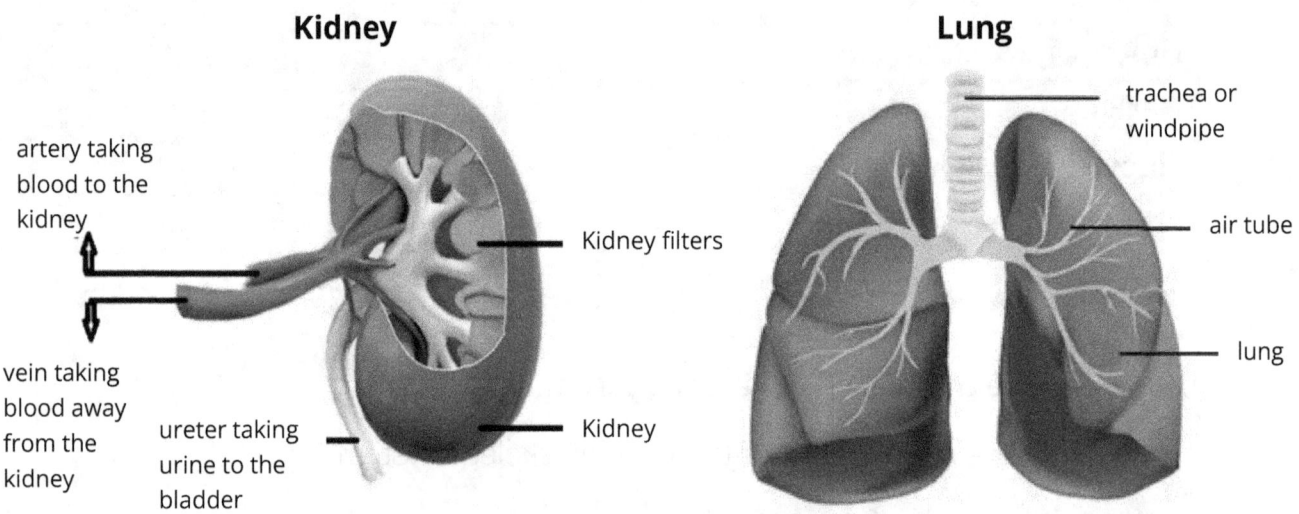

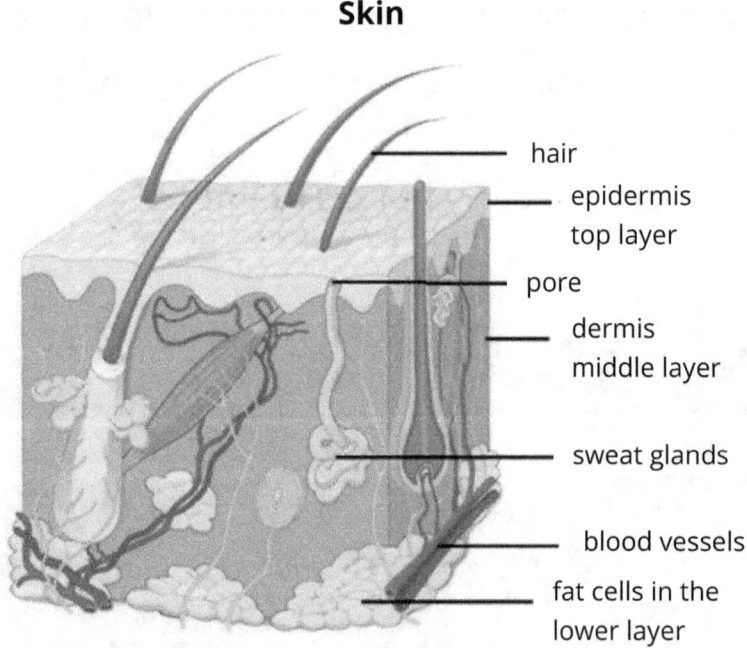

Figure 8.10: *Organs in the excretory system.*

Parts of the excretory system

Like the digestive system, the excretory system has many parts that help it perform its functions.

Kidney: The kidney is a bean-shaped organ that filters out waste, called urine, out of the body. The kidneys remove waste products from the workings of all the organ systems. The human body has two kidneys that pass urine to the bladder, which then sends the urine out of the body through the urethra.

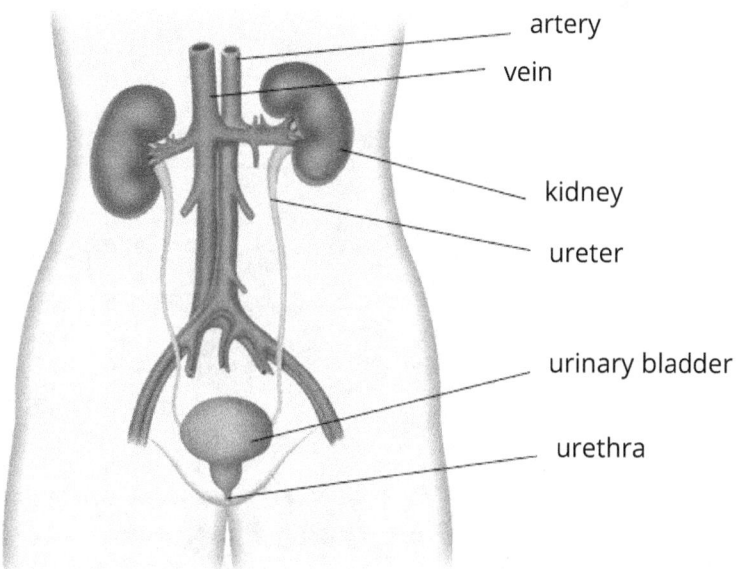

Figure 8.11: The human excretory organs - urinary system.

Lungs: The human body has two lungs. When you exhale or breathe out, the lungs remove carbon dioxide and water from the body.

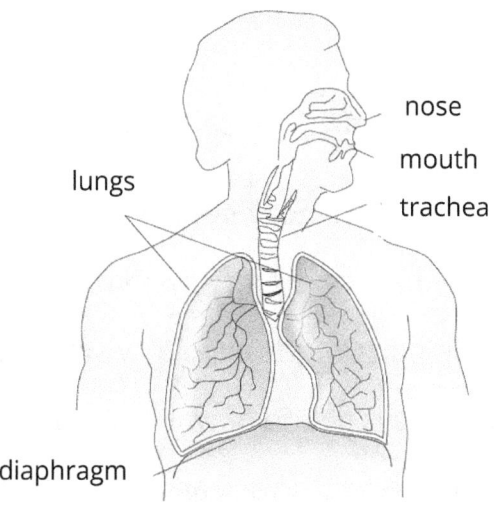

Figure 8.12: The human body lungs.

Skin: The skin is the largest organ of the body, and it covers the entire body. The skin may look thin, but it is divided into several layers, as shown in Figure 8.13. Each layer supports the skin's many functions (such as removing urea, salt and water from the body).

The skin contains special parts called sweat glands. Through the sweat glands, the skin removes extra water and salt (sweat) through tiny holes in the skin called *pores*. The skin contains sensitive cells, which detect pressure, temperature and pain. The skin also has oil glands to keep it moist.

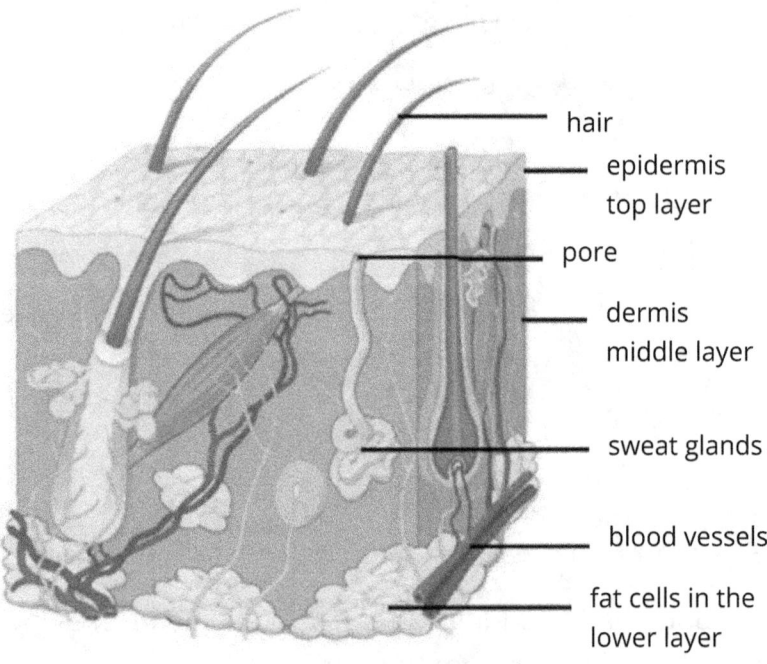

Figure 8.13: The human skin.

Let's take another look at the excretory system.

1. What is the main function of the excretory system?

2. What are the THREE main parts of the excretory system?

3. Which organs help the kidneys get rid of water and salt?

4. What are the main waste products in sweat?

5. What would happen if the excretory system does not function properly?

Herman Biggs once said, 'The human body is the only machine for which there are no spare parts.'

With advances in technology, is this saying true? What do you think? You may have to do online research.

Activity 8.4.1: Information and communication technology (ICT): The excretory system

Using simple materials, make a model of the excretory system that is the urinary system: kidneys, bladder and ureters. Outline the human body on a piece of stiff paper or cardboard. Use materials such as red peas, spaghetti, glue, mini-marshmallows, drinking straws, cord or string.

Activity 8.4.2: Word find

In this puzzle, you will find the keywords used in Chapter 8. Find and circle the listed words found in the puzzle.

- locomotion
- skeleton
- bone
- muscle
- joint
- digestion
- alimentary canal
- oesophagus
- stomach
- intestine
- anus
- lungs
- skin
- kidneys
- sweat
- urine
- urea
- ureter
- bladder
- pancreas

B	O	N	E	R	O	N	M	U	S	C	M	U	S	C	L	E
L	O	E	S	O	P	H	A	G	U	S	W	A	M	P	O	P
A	L	I	M	E	N	T	A	R	Y	C	A	N	A	L	C	A
D	U	N	S	T	O	M	A	C	H	M	Z	X	C	A	O	N
D	N	T	A	D	A	P	T	A	T	I	O	N	A	S	M	C
E	G	E	B	N	N	M	Q	W	E	D	A	Y	U	I	O	R
R	S	S	K	E	L	E	T	O	N	I	K	I	E	E	T	E
E	I	T	C	J	V	I	P	M	A	G	I	I	D	S	I	A
T	C	I	I	O	A	V	R	D	S	E	O	C	D	S	O	S
E	O	N	N	I	S	N	A	I	D	S	S	E	R	N	N	E
R	U	E	D	N	A	P	U	K	E	T	K	I	D	N	E	E
U	R	E	A	T	L	A	I	S	T	I	I	R	R	E	T	Y
H	I	H	Y	S	I	C	A	L	F	O	N	T	U	R	E	S
P	N	E	S	W	E	S	A	T	I	N	S	W	E	A	T	J
W	E	I	N	T	E	S	T	I	N	E	S	S	O	I	V	E

Evaluate yourself!

Use this evaluation grid to check your understanding of the concepts discussed in this chapter. Read each statement below and insert the symbol that best shows how well you feel you understand the concept. Ask a teacher or parent to help you go over any areas that are still unclear, or that you do not feel you have mastered. Be honest!

I got it!

I need to do more work.

I do not get it. I need help.

In this chapter:

	I got it!	I need to do more work.	I do not get it. I need help.
1. I can explain what is meant by the term system.			
2. I can identify the organ systems and state their functions in humans.			
3. I can recognise the integration of the different organ systems in carrying out life processes.			
4. I can identify selected organs in the human digestive system (mouth, oesophagus, stomach, small intestine and large			

	I got it!	I need to do more work.	I do not get it. I need help.
intestine) and outline the path food travels from mouth to anus.			
5. I can describe how the bones, joints and muscles work together to produce movement in humans.			
6. I can identify the excretory organs of humans (kidneys, lungs and skin) and state their role in excretion.			

Grab your Merge Cube to explore the augmented reality activities next.

Augmented reality: Mr body

We'll look at some stylised human anatomy on this card. Anatomy is the study of the biological structures of humans and other animals. Dissecting these organisms is what helps us define their structure.

Activity: Mr body

In this experience, tap on individual organs to expand them and see more information. Take your time to explore. Read the pop-up information panels about all the objects you come across. Which organ helps us circulate our blood?

> Do you know you are capable of doing anything you put your mind to? As Elliot Hulse said, you can 'become the strongest version of yourself.'

> Let's go, let's go! Let's jump to the next chapter and leap for some more knowledge.

TERM 2, Unit 3:
Mixtures and separation techniques

Words worth knowing

The list below contains the scientific words that will be discussed in this unit.

- colloid
- compound
- distillation
- evaporation
- filtration
- filter
- decanting
- mixture
- sieving
- solution
- suspension

When life gives you mixtures, use your science skills to separate and conquer. From sieving to filtering, there's always a smart way to sort things out!

Chapter 9:
Mixtures and separation techniques

Chapter objectives
Specific objectives:
- ✓ Demonstrate that a mixture is made up of two or more substances.
- ✓ Classify mixtures as solutions, suspensions and colloids.
- ✓ Recognise that all mixtures can be separated.
- ✓ Demonstrate the separation of selected types of mixtures using various techniques.
- ✓ Conduct investigations with due regard to safety.
- ✓ Use appropriate scientific vocabulary to describe mixtures

CHEETAH Science Fiction

Saved by Science

The kingdom of Pernassa was in trouble. The vile legion of Carcassius had contaminated the Rogue River, the lifeblood of the kingdom, with a toxic substance so that all who drank from it turned into green lizards.

The king decided to do something about it. And the fastest way to get it done was to organise a competition called The Cleansing of the Rogue. The prize? Anything the heart desires. Three brave citizens took the king's challenge.

First up was Mrs Witherspoon. She brewed a deadly-smelling potion, spoke to the wind, and whispered to the animals of the forests. She did a little dance and then spoke to the king, who was waiting and watching on a tall throne by the side of the river.

'It is done, your majesty!'

'Then drink from the river and you'll have anything your heart desires!' bellowed the king.

'Umm, drink from the river?' she asked in disbelief. She had not expected that. 'Okay, then.'

Mrs Witherspoon took a small sip from the river and before she could even close her lips, she transformed into a big, green lizard.

'Next!' the king shouted. Zoon, infamous harbinger of mischief, looked at the new lizard hopping around aimlessly on the riverbank and decided to forfeit. The king rolled his eyes and looked at the last hope of the kingdom of Pernassa—Timon. Timon the peasant. Timon, the town's smart man, who was always experimenting, said he had a solution to the problem. Timon, who had a new practice called 'science'. The king was not sure how to pronounce it. 'Sianse?'

Timon walked slowly to his experiment, which he had set up early that morning, removed the cloth covering, swiping it off dramatically. The king and the onlookers oohed and aahed with amazement when they saw what was underneath. It was a strange collection of bowls and pipes and bubbling liquid over a flame.

'Behold King of Pernassa! I give you distillation!' Timon's voice echoed. And before the king could even ask, Timon poured liquid from the large apparatus into a cup and drank from it. The crowd was silent for a few seconds and then burst into applause.

'Timon! Timon! Timon!' the crowd cheered.

'May I have that wish now?' Timon asked.

9.1: What is a mixture?

Figure 9.1: *A mixture of candies, raisins and nuts.*

Look around your classroom! The classroom is a mixture of males and females. At some time during the class, the teacher will separate this mixture into those who came early and those who came late. Go to the library. There is a mixture of all types of books, which can be separated according to the different authors.

The world around us contains many types of **mixtures**. A mixture is made up of two or more substances. In a mixture, the combined substances do not form a new substance; even when combined, each substance in a mixture remains the same. This means that mixtures can be separated back into individual substances. Have you ever had a large glass of lemonade? You have had a mixture! Lemonade is a mixture of lemon juice, water and sugar. Mixtures can contain an even or uneven mix of the components that make them up. Candies with nuts are a mixture as the candy and nuts will remain as they were before they were mixed.

Let's examine the properties of mixtures.

1. What is a mixture?
2. Give an example of a mixture that you use every day.
3. Would combining sugar and flour make a mixture?
4. Would Jamaican rice and peas dish be considered a mixture?
5. Is water a mixture?

9.2: Classifying mixtures

Types of mixtures

There are many different ways of making mixtures. Mixtures can be made between two or more solids; between two or more liquids; or between two or more solids and liquids.

Substances	Examples
Solid-solid	pens and pencils
Solid-liquid	sugar in alcohol
Liquid-liquid	oil in gasoline

Activity 9.1.1: Observing and grouping mixtures

Work in groups to observe and group various mixtures.

Materials:

- individual, clear containers, each holding the following 12 mixtures:
    - alcohol and water
    - soil and oil
    - sand and water
    - salt and water
    - oil and alcohol
    - spoons and forks
    - nails and water
    - salt, sand and nails
    - sand and paperclips
    - vinegar and water
    - oil and water
    - pencils and pens

Procedure:

1. The teacher will set up twelve workstations and number them from 1 to 12. Each workstation will be provided with one container of each mixture. Each mixture should be labelled based on the substance in the mixture.

2. Students will be placed in groups of THREE. Each group is to be seated at a workstation.

3. Each group will discuss and sort the types of mixtures at each workstation into solid and solid, solid and liquid or liquid and liquid.

4. Groups should place the results on the table. An example is provided below.

Station #	Type of substance	Mixture group
1	pants and shirts	solid and solid

5. Groups will move to the next station when indicated by the teacher.

6. The class will have a teacher-led discussion after each group has visited all twelve stations.

Solutions, colloids and suspensions

Scientists use their observation of liquid mixtures to group some mixtures as solutions, colloids or suspensions.

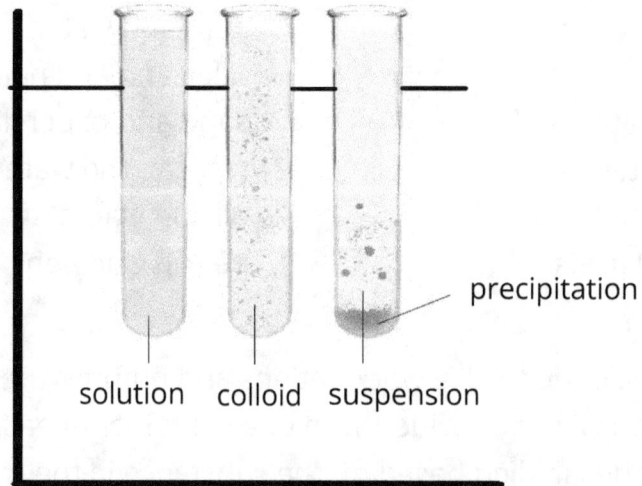

Figure 9.2: Solution, colloids and suspensions.

A **solution** is a mixture that has one or more substances (**solutes**) dissolved in another substance (a **solvent**). In a solution, the solute particles do not settle to the bottom of the container when allowed to stand. In a solution, the solute particles cannot be filtered out.

Saltwater is an example of a solution. The salt particles cannot be seen and will not settle to the bottom of the glass over time.

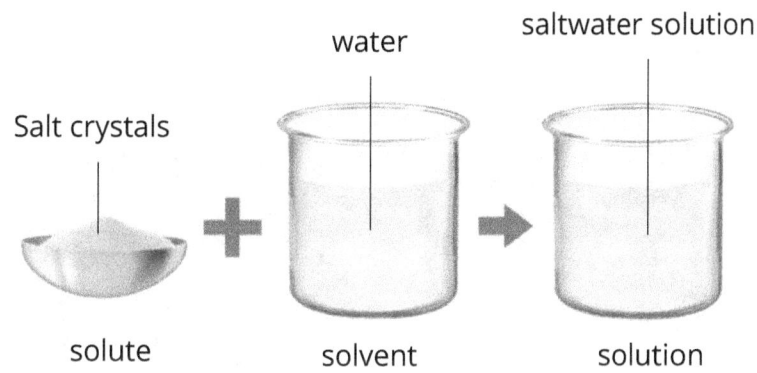

Figure 9.3: *Salt and water solution.*

A **suspension** has particles that are large enough to be seen by the naked eye. These particles settle to the bottom of the container over time if left undisturbed. Sand or soil in water and dust in the air are all examples of suspensions.

If left unshaken, the particles will settle to the bottom. For example, milk of magnesia given by doctors to help with acid indigestion will settle at the bottom of the bottle over time. The bottle of the medicine will say 'Shake well before taking'.

A **colloid** has medium-sized particles, somewhere between a solution and a suspension. Colloids will not naturally settle over time like suspension. Examples of colloids include milk, toothpaste, paint, mayonnaise, fog, creams and gels.

Did you know that smoke is a mixture of particles suspended in the air?

Activity 9.2.2: Does it dissolve?

Work in groups of THREE to investigate which solids and liquids dissolve.

Materials:

- five plastic bottles (500 ml each) with covers
- five teaspoons
- water
- samples: chalk, salt, flour, sugar and sand

Procedure:

1. Label the five plastic bottles: chalk, salt, flour, sugar and sand.

2. Add one teaspoon of each sample to the empty bottle, as labelled.

3. Add water to half the bottle.

4. Cover the bottle, shake vigorously and then allow the bottles to settle.

5. Record your observations after one day. Did the sample dissolve in the water? Did the sample settle upon standing?

6. Make a chart like the one shown below to record your findings.

Mixture	Observation	Dissolved? (Y/N)
e.g. chalk + water		

Extend your learning

Try leaving some of your samples on a sunny window for a few days. Has anything changed? Can you explain why?

Let's take another look at classifying mixtures.

1. What is a solution? Give an example.

2. What is a suspension? Give an example.

3. What is a colloid? Give an example.

4. Give one similarity between a sugar solution and a salt solution.

5. You made a mixture of sugar and water, and you noticed some solids at the bottom of the cup. Does that mean that sugar does not dissolve in water?

9.3: Separating mixtures

You learned in the previous section that a mixture is made up of two or more substances. Sometimes it is possible to separate a mixture by hand (sorting), but often extra equipment is needed to carry out a more difficult separation. Filtering, decanting, sieving, evaporation or using a magnet are some of the more common methods for separating mixtures.

Filtering

Filtration is the process of separating a mixture of a liquid and solids, which do not dissolve in the liquid. Filtering is used to separate solid particles from a liquid. For example, filtration can separate sand and water or the coffee grounds from the water.

When filtering, the mixture is usually poured into a funnel containing filter paper. The filter paper traps the larger particles of the solid because they are too big to pass through its pores while allowing the liquid particles to pass through.

Figure 9.4: *Filtration of mixtures.*

Did you know separation techniques are used to produce common food items like salt from the ocean or juice from a fruit?

Filtration is one of the steps used by the National Water Commission to purify water from the dams and reservoirs before it is sent to your home.

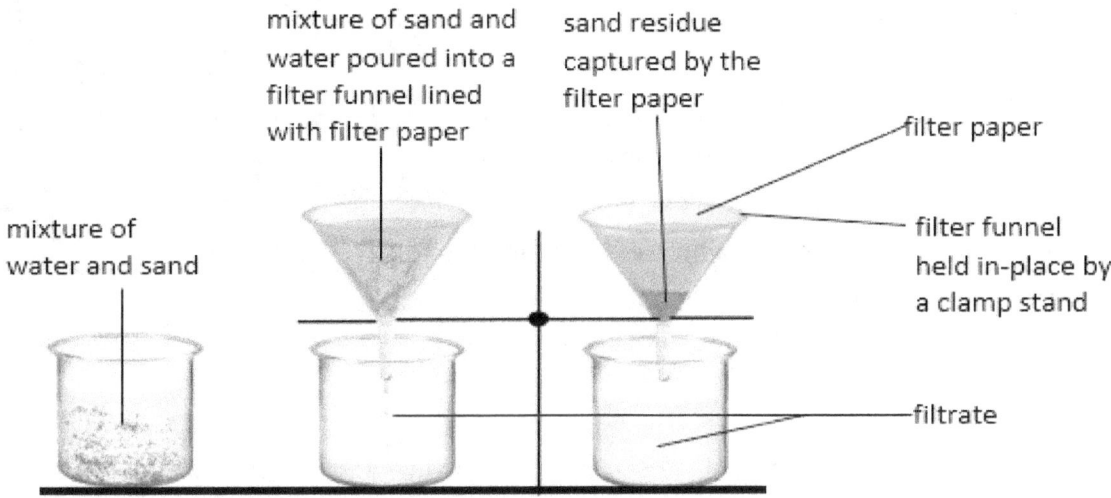

Figure 9.5: Example of filtration (separation of sand from water).

The separation technique we use depends on the properties of the components of the mixture.

Decanting

Decanting is the separation of a mixture by gently pouring off the top layer of two liquids that do not mix, such as oil and water. *Note the method of pouring the liquid alongside the straw to prevent spills.*

Figure 9.6: Decanting of liquids that do not mix, such as water and oil.

Sieving

If a mixture is made of solid particles of different sizes, it can be separated by sieving. The larger particles will remain behind in the sieve, while the smaller particles pass through the holes. You may have used a strainer to sieve flour or cornmeal to be more refined by removing the larger sizes. You may have seen someone working on a construction project using a sieve to separate sand from gravel.

Figure 9.7: *Sieving flour.*

Evaporation

Evaporation is a way of separating solids from liquids using heat. When most of the liquid has evaporated, the mixture is removed from the flame and allowed to cool. As the mixture cools, the last of the liquid evaporates, leaving the solids in the evaporating dish. One example is a salt solution.

Figure 9.8: *Evaporation.*

Did you know that painting uses the separation process of evaporation? The wet paint is a mixture of colour pigment and a solvent. When the solvent dries and evaporates, only the colour pigment is left?

Magnets

Magnets can remove non-magnetic materials from magnetic materials in a mixture. This is usually done at a junkyard, especially during the recycling process.

Figure 9.9: *Junkyard magnet used to separate large quantities of metal from non-metals.*

Mixtures are an important part of everyday life. We need to be able to separate mixtures into their parts. This allows us to use, reuse and recycle their components.

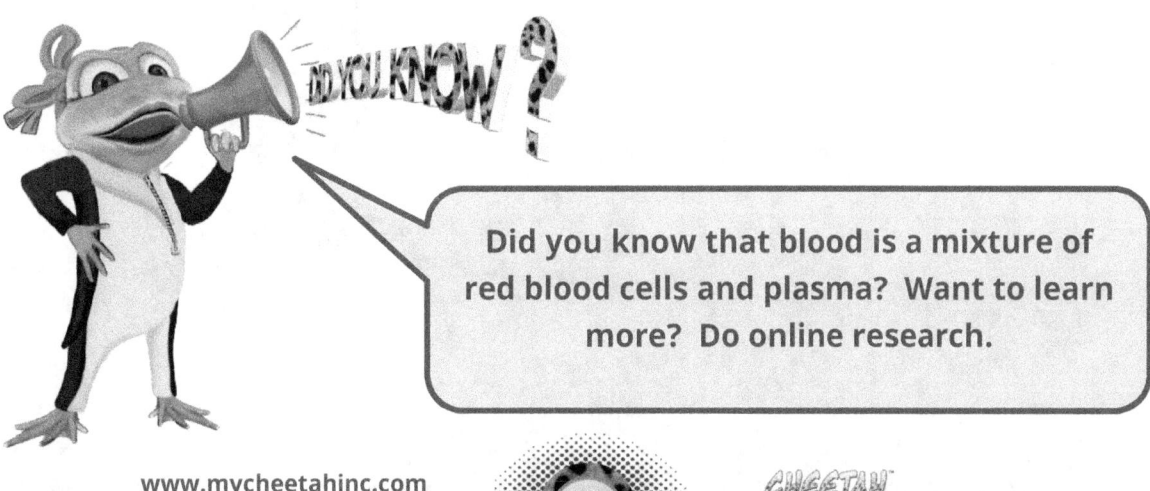

Did you know that blood is a mixture of red blood cells and plasma? Want to learn more? Do online research.

Activity 9.3.1: Using magnets to separate

Materials:

- a variety of mixtures of metals and non-metals, some of which can be separated with a magnet
    - Mixture 1: small steel nails and rice, labelled M
    - Mixture 2: small steel nails and brass/aluminum nails, labelled N
    - Mixture 3: metal paperclips and sawdust, labelled O
    - Mixture 4: paper, sand and salt, labelled P
- a small magnet
- plastic dishes or plates to hold mixtures

Procedure:

1. Collect one of the sample mixtures in a dish.
2. Hold the magnet over the mixture.
3. Record your observations.
4. Repeat steps one to three with the other sample mixtures.

Activity 9.3.2: A separation challenge

Work in groups of three to separate the mixture into all its separate parts.

ATTENTION
Adult supervision required for this activity.

Materials:

- a funnel and filter paper
- a sieve
- a magnet
- evaporating dish (glass dish)
- a mixture of salt, sawdust, paper clips and gravel
- water
- heat source (e.g. spirit lamp, candle and Bunsen burner)
- paper

Procedure:

1. Discuss the best order to separate the mixture into all of its parts and formulate a plan.
2. Carry out all procedures necessary to present each part of the mixture on individual pieces of paper.

HINT: You may need to add something to the mixture to complete the challenge.

Let's examine the methods of separating mixtures.

1. Describe THREE common ways to separate mixtures.
2. Explain ONE way to separate sand from water.
3. Why are separation techniques useful?
4. What is meant by decanting and how can we use it in the home?
5. You have a mixture of oil and water. Tell a friend how to separate them.

CHEETAH® PREPARING THE JAMAICAN SCIENTIST

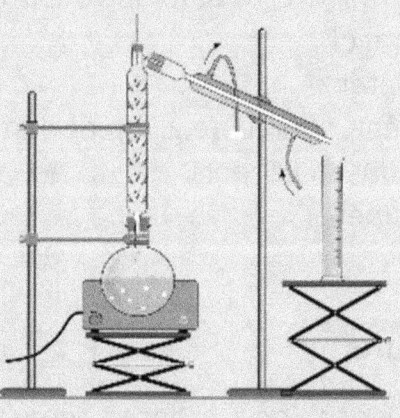

Did you know fractional distillation can be used to separate liquid mixtures with different boiling points? For example, alcohol mixed with water.

Activity 9.3.3: Information and communication technology (ICT): Separation of mixture

Every year, Jamaica buys crude oil (petroleum) in bulk. We then separate this mixture to get various types of oils. For instance, cooking gas, gasoline, diesel oil, kerosene oil, engine oil, asphalt and hair oil (petroleum jelly).

Conduct research to find out how crude oil as a mixture is separated into all the various substances. Share your findings with the rest of the class.

Activity 9.3.4: Can you locate the right apparatus?

Look at the picture below and list the separation methods observed.

Evaluate yourself!

Use this evaluation grid to check your understanding of the concepts discussed in this chapter. Read each statement below and insert the symbol that best shows how well you feel you understand the concept. Ask a teacher or parent to help you go over any areas that are still unclear, or that you do not feel you have mastered. Be honest!

I got it!

I need to do more work.

I do not get it. I need help.

In this chapter:

	I got it!	I need to do more work.	I do not get it. I need help.
1. I can show that a mixture is made up of two or more substances.			
2. I can group mixtures as solutions, suspensions and colloids.			
3. I understand that all mixtures can be separated.			
4. I can show the separation of selected types of mixtures using various techniques.			
5. I can do investigations with due regard to safety.			
6. I can use appropriate scientific vocabulary to describe mixtures.			

CHEETAH® PREPARING THE JAMAICAN SCIENTIST

How eager are you to learn? According to Brian Hibbert, 'The capacity to learn is a gift; the ability to learn is a skill; the willingness to learn is a choice.'

Mek wi hop to di next unit.

Let's go, let's go, let's leap for knowledge!

TERM 3, Unit 1:
Diet and food

Words worth knowing

The list below contains the scientific words that will be used within this unit.

- balanced diet
- calories
- carbohydrates
- constipation
- dairy products
- diabetes
- diet
- fats
- food groups
- fruits
- grains
- insulin
- macronutrients
- micronutrients
- minerals
- nutrients
- nutrition labels
- obesity
- proteins
- saturated fats
- type 1 diabetes
- type 2 diabetes
- unsaturated fats
- vegetables
- vitamins

Share this quote by Hippocrates with your relatives:

'Let food be thy medicine and medicine be thy food.'

Chapter 10:
Healthy eating habits

Chapter objectives

- ✓ Explain some of the consequences of not having a balanced diet.
- ✓ Assess the causes of obesity, diabetes and malnutrition.
- ✓ Outline measures to mitigate against selected lifestyle diseases.
- ✓ Justify the need for eating healthy foods.
- ✓ Evaluate data to draw conclusions about the consequences of improper diets.
- ✓ Show concern for others who make unhealthy eating choices.
- ✓ Show sensitivity to individuals who suffer from food-related illnesses or challenges.
- ✓ Use appropriate scientific language related to food and health.

CHEETAH Science Fiction

The Race

'I want to eat cake all day long!' said LaChase the Cheetah, stuffing a slice of cream pie into his mouth. 'I have a race to run next month.'

'That wouldn't be wise if you don't exercise regularly,' said Hooty Hoot, the wise owl. 'You've got a race next month, so you need to eat healthy food. It will give you the energy and strength that your body needs to win.'

LaChase opened the kitchen cupboard and stared at the loaf of freshly baked banana bread. 'What's food got to do with it?' he said. 'I can beat anyone in a running contest, anyone on the planet! Next month's race is just a formality. I've already won.'

Hooty Hoot didn't like it when LaChase boasted. 'No one likes a show-off. You'd better exercise,' she said, as LaChase gobbled the sweet treat.

A week before the race, LaChase noticed that he had put on a few pounds, but still felt confident that he would win the race. On the day of the race, Hooty Hoot noticed that everyone appeared fit and ready to race — everyone except LaChase.

The starting pistol was fired and Cheetah launched into an early lead. However, by the time he had reached the first corner, a rival was already gaining on him.

'What's wrong with LaChase?' asked Croaky Croak, watching the race at the side of the track with Hooty Hoot. 'He doesn't look so fast today.' One minute later, LaChase stumbled over the finish line. He was in the fifth position, the worst position he had ever finished in. In fact, until today, he had won every race he had competed in.

On the way home, walking with Croaky Croak and Hooty Hoot, LaChase was still upset with the result. 'The other contestants must have cheated!' he said. 'I NEVER LOSE A RACE!'

'What have you been eating this last week?' asked the owl. 'I told you to exercise. This is what happens when you eat and don't exercise. You put on weight.' LaChase was silent for the rest of the journey home.

10.1: Balanced diet and how we get to it

John went to school without breakfast and started to feel dizzy in class. His teacher asked what was wrong and then sent him to the canteen to get some food. He got a cup of cornmeal porridge, an egg sandwich and a cup of mint tea. John felt much better and returned to class happy and focused, but did John eat a balanced diet?

10.2: What is a balanced diet?

What is a balanced diet?

A diet is the type of food and drinks that a person consumes. Having a balanced diet means eating from all the different food groups in the right proportion depending on the age, physical activity and health of the individual.

Figure 10.1: *food groups*

For example, older people need more protein than carbohydrates because they are repairing tissues and they do not need so much energy. Younger people also need carbohydrates to give them energy for their increased activity, and protein for building tissues and more rapid physical growth. Older people need more calcium as they lose bone density. Children also need more calcium as they are growing and building larger bones.

There are six main food groups: staples, legumes, vegetables, fruits, fats and oils and foods from animals. To stay healthy, the human body needs to get different types of nutrients from these food groups. But it is also important to eat the food to get the nutrients in the right proportions to maintain good health.

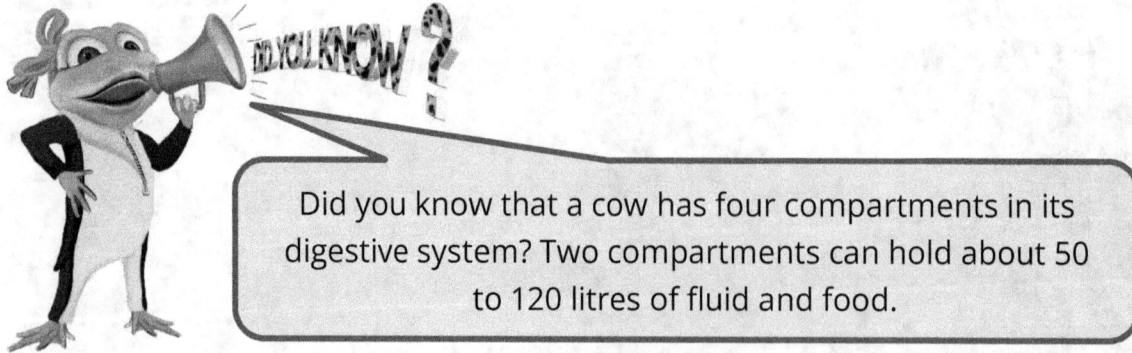

Did you know that a cow has four compartments in its digestive system? Two compartments can hold about 50 to 120 litres of fluid and food.

Figure 10.2: food based dietary guidelines for Jamaica

Adapted from the Ministry of Health and Wellness, Jamaica for educational purposes.

Food groups

Here are some sources and functions of food groups:

Staples are high in carbohydrates and low in fat, such as rice, sweet potato, Irish potato, yam, flour, banana, oats, cornmeal, yam, bread and breadfruit. These are energy foods.

Legumes are rich in protein and oils along with smaller amounts of carbohydrates. Examples include corn, nuts, grains, peas and beans.

Foods from animals are rich in protein and oils, such as beef, chicken, pork, fish, mutton, and milk. These foods have many uses but are most helpful in the repairing of tissues. Dairy, including milk, cheese, and yogurt, is made from animal products.

Fruits are plant products that are low in fat and calories and include foods like oranges, bananas and strawberries. They are a source of vitamins, minerals, fibre and water. Eating pomegranate prevents hair loss, as it strengthens the hair. Eating fruits also prevents constipation and helps the hair, skin and nails to grow.

Vegetables are plant products that have high amounts of vitamins and minerals. They include foods like spinach, callaloo, cucumbers and broccoli. Vegetables are a good source of fibre and help to prevent constipation. Eating lots of vegetables can also help to control diseases and conditions like obesity, high blood pressure and diabetes.

Fats and oils give energy and form a protective covering for organs. Fats are contained in foods such as butter, nuts, avocado and ackee, while oils are made from vegetables and seeds such as coconut, olive, sunflower and soya.

10.3: Consequences of not having a balanced diet

You have learned that eating healthy foods can keep your body strong and healthy. The opposite is also true. Not eating a balanced diet can lead to health problems such as malnutrition, diabetes, obesity and constipation.

Malnutrition is caused when a person is not eating the right amounts and types of food. Malnutrition can be caused by not eating enough calories, eating too many calories or eating incorrect amounts of nutrients.

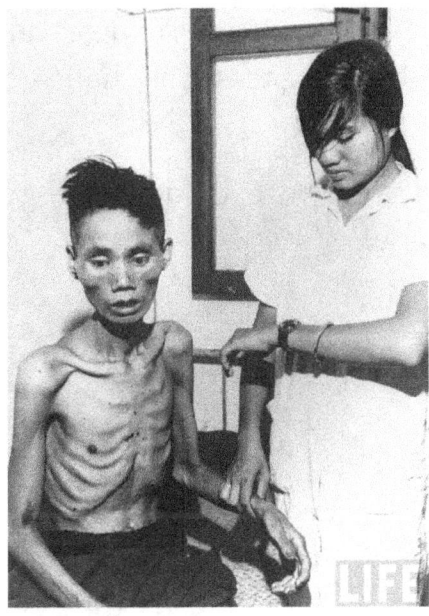

Figure 10.3: *Man starved of food.*

Adapted from wordpress.org for educational purposes.

`1966 Le Van Than, a 23-year-old Nationalist who defected from Communist forces, was recaptured by the Viet Cong and deliberately starved for 1 month in their prison camp by Manhhai. Licensed under CC BY 2.0.

People who are malnourished are often underweight. Malnutrition can make it difficult to concentrate. That makes it hard to do well in school. Malnourished people also have lower physical performance in sporting activities because they do not have the energy from the foods they eat to grow and function. Some of the causes of malnutrition include poverty, ignorance, droughts, wars and plagues.

What is diabetes?

People with diabetes have high amounts of sugar in their blood. There are two types of diabetes: Type 1 diabetes and Type 2 diabetes.

Type 1 diabetes usually develops in childhood or adolescence and cannot be prevented. In this type of diabetes, the body produces little to no insulin. Insulin controls the amount of sugar in the blood, so, without it, the body loses control of its sugar levels.

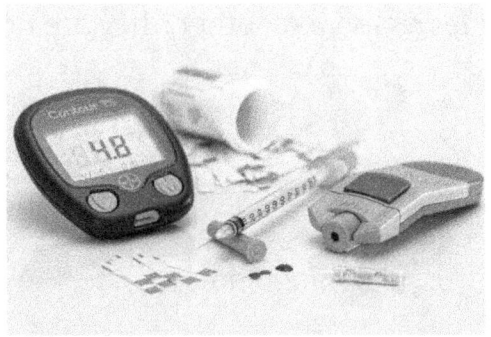

Figure 10.4: *Medical equipment used to monitor diabetes.*

Type 2 diabetes usually develops in adulthood if a person is overweight. In this type of diabetes, the cells do not respond to insulin as well as they should, or the body may not produce enough insulin to control the levels of sugar in the blood.

Both types can increase the chances of having other health problems, like heart and kidney disease. Having a family member with diabetes increases the chance of developing it.

Diabetes can be controlled by injecting insulin into the blood each day, eating a balanced diet that is high in fibre and exercising regularly.

Obesity can be caused by an unhealthy diet. People who are obese are over their recommended weight. Obesity can be caused by an inactive lifestyle and making poor food choices. Obesity increases the risk of heart disease, high blood pressure and diabetes.

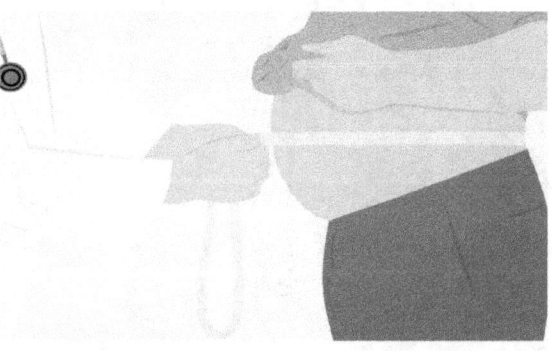

Figure 10.5: Waist measurement.

Constipation happens when you have a hard time having a bowel movement when a person goes to the toilet. It can be caused by not eating enough dietary fibre. Constipation can be uncomfortable. Increasing the number of whole grains, fruits and vegetables and drinking plenty of water can reduce the risk of constipation.

Prevention of lifestyle diseases

Lifestyle diseases are based on high-stress levels and daily habits such as lack of exercise, smoking and poor food choices. These diseases cannot be transferred from one person to another; they are not contagious. Examples are:

1. stroke
2. heart disease
3. diabetes
4. obesity
5. high blood pressure.

Figure 10.6: Jamaica's Ministry of Health and Wellness encourages healthy lifestyle choices.

Adapted for educational purposes: Photo by Rudranath Fraser, 2017 – Jamaica's Minister of Health, Dr. Christopher Tufton, has initiated the campaign `Jamaica Moves' to encourage citizens to make healthy lifestyle choices.

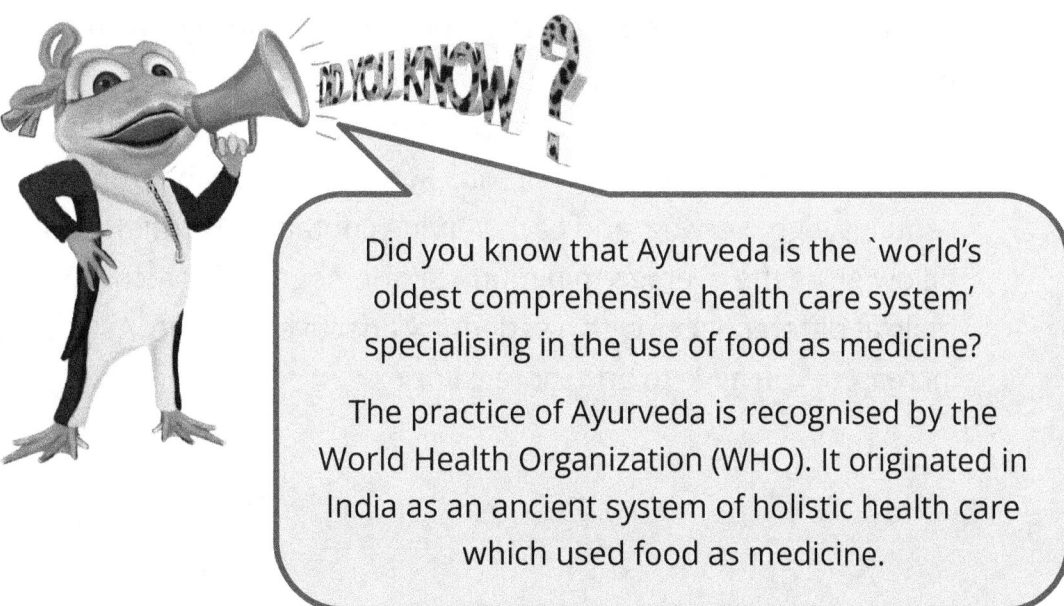

Did you know that Ayurveda is the `world's oldest comprehensive health care system' specialising in the use of food as medicine?

The practice of Ayurveda is recognised by the World Health Organization (WHO). It originated in India as an ancient system of holistic health care which used food as medicine.

There are several ways to prevent these lifestyle diseases, such as:

1. eating a well-balanced diet, taking in vitamins and minerals and not consuming too much fat, carbohydrates and salt
2. exercising regularly will help a person burn excess calories and lose weight
3. avoiding too many alcoholic beverages
4. getting more sleep
5. making more informed choices of packaged and processed food by reading nutrition labels. Packaged foods are required to have nutritional information printed on them.

Nutrition labels give information on the food we are eating, such as:

Figure 10.7: *nutrition facts*

- ✓ the serving size
- ✓ how to store the item and the period when the product can be safely consumed (product's shelf-life). Shelf-life is usually written as 'best before' with an expiration date
- ✓ the ingredients and the calories in the food. Calories are a measurement of the amount of energy in food. See from the label, one container contains eight serving, and each serving contains 230 calories. The table below gives the average minimum calorie needed for a child, female or male at different ages with a sedentary (inactive) lifestyle. As your activity increases, you have to add more calories.

Table 10.3: Calories needed by people of different genders and ages who do not live active lives (sedentary)

Gender	Age (years)	Sedentary (Calories)
Child	2-3	1,000
Female	4-8	1,200
	8-13	1,600
	14-18	1,800
	19-30	2,000
	31-50	1,800
	51+	1,600
Male	4-8	1,400
	8-13	1,800
	14-18	2,200
	19-30	2,400
	31-50	2,200
	51+	2,000

Did you know that exercise prevents cardiovascular (heart) disease?

Let's examine the effects of not having a balanced diet.

1. What are some reasons why people develop nutritional disorders?
2. Choose one of the nutritional disorders. Explain what it is and how it can be prevented.
3. If the body feels weak and tired, what type of food could be recommended?
4. If an individual has had a cut that will not heal for longer than expected, what group of nutrients could be recommended?
5. A patient's eyesight is failing due to a vitamin deficiency. Conduct research to find out which vitamin should be included in the diet.

Some reasons for unhealthy food choices

Fresh fruits, vegetables and other healthy foods can be very costly to obtain. In some places, people do not have easy access to food, clean water and suitable living conditions. As a result, people suffer from illnesses caused by poor diet and unclean water that could be prevented.

According to the United Nations' recent estimates, `10 percent of the world's population or 734 million people lived on less than USD 1.90 a day.' How much is that in Jamaican currency?

Activity 10.3.1: Information and communication technology (ICT): Poster competition

Research and create an electronic poster in support of the Ministry of Health and Wellness (MOHW) campaign 'Jamaica Moves'.

Share your poster in your class online group (google classroom) so that your classmates can vote and select one poster to represent the class. Submit this poster to the MOHW requesting that your class's selection be included in this campaign.

Activity 10.3.2: Unscramble the puzzle

Unscramble the following using the clues.

1. siebyto
2. Ttaiimnoalur
3. stiaeedb
4. tvcyus
5. eristkc

Clues

1. A medical condition in which a person is carrying more body weight than is recommended
2. A medical condition caused by an inadequate diet, such as by eating too few calories, too many calories, or not enough necessary nutrients
3. A medical condition that causes high blood sugar levels; is caused when the body does not produce enough insulin to remove sugar from the blood or the cells do not respond to insulin
4. Vitamin C deficiency leads to weakness, anemia and bleeding gums disease
5. Vitamin D deficiency leads to soft weak bones

Evaluate yourself!

Use this evaluation grid to check your understanding of the concepts discussed in this chapter. Read each statement below and insert the symbol that best shows how well you feel you understand the concept. Ask a teacher or parent to help you go over any areas that are still unclear, or that you do not feel you have mastered. Be honest!

I got it!

I need to do more work.

I do not get it. I need help.

In this chapter:

	I got it!	I need to do more work.	I do not get it. I need help.
1. I can explain some of the consequences of not having a balanced diet.			
2. I can assess the causes of obesity, diabetes and malnutrition.			
3. I can outline measures to mitigate selected lifestyle diseases.			
4. I can justify the need for eating healthy foods.			
5. I can evaluate data to draw conclusions about the consequences of improper diets.			
6. I can show concern for others who make unhealthy eating choices.			

	I got it!	I need to do more work.	I do not get it. I need help.
7. I can show sensitivity to individuals who suffer from food-related illnesses or challenges.			
8. I can use appropriate scientific language related to food and health.			

> Today is a good day and a gift that will only last 24 hours. What will you do with this present that was wrapped for you?

Chapter 11: Classification and effects of drugs

Chapter objectives

- ✓ State the meaning of the term 'drug'.
- ✓ Classify commonly used drugs as legal or illegal.
- ✓ Explain the importance of following guidelines on the proper use of a drug.
- ✓ Describe the effects of drugs on the body.
- ✓ Show responsible behaviour in the use of drugs.

CHEETAH Science Fiction

Ice Lolly

It all started the day she had the delicious ice pops. Joel bought them across the street from her school playground from a funny-looking young man in a bright orange tee, luminous green sneakers and a constant grin. Joel, who sat behind her in math class, had whispered about him.

'I'm telling you, Samantha, you will freak out! He sells them for only a dollar!' Samantha was already curious.

'If you're buying...' she retorted a moment later. Samantha could feel Joel's grin.

'I'll get you the blue one, my favourite! Better than any ice cream you've had!' Joel's excitement almost made him vibrate in his seat.

'Fine! You know where I always sit in the playground. Bring it there. I'm not sneaking through the fence with you!' Samantha always played at being bothered by Joel when she needed something from him. It worked every time. In less than five minutes after the recess bell rang, Samantha enjoyed a heavenly treat. 'WHAT IS THIS?' Shocked by Joel's discovery, she was already blown away by how sweet the ice lolly was after just a few licks. Her taste buds were screaming with delight. More than that, she suddenly felt very alive. She had hit the pleasure jackpot!

'I told you, this stuff is nuts! I'm getting us two more!' Joel sucked on his almost finished lolly before bounding for the school fence. He could not care less about the streaks of blue liquid dribbling down his chin. That day, he and Samantha would eat six lollies each, bought from the guy in the orange tee from across the school playground. Increasingly, Samantha's head felt light and it was hard to stay awake in class.

Samantha and Joel grew their friendship with the 'kisko guy' as they referred to him. Samantha grew suspicious when he offered them what he called 'yummy bears', bear-shaped pills, bright blue and each the size of two candies. 'Are these drugs? If they are, I'm not taking any, and neither should you, Joel' Samantha could not shake the weird feeling in her gut.

'They will make you clever, and I know how you are struggling with math, Joel. How is that a bad thing? Samantha?' 'kisko guy' winked at them, his charm ever-present. All they had been taught about drugs by their parents and at school went out the window.

It was only a matter of weeks before Joel and Samantha couldn't go a day without the 'yummy bears'. A year later, both teens were school dropouts. Samantha was in a juvenile drug rehabilitation centre in her hometown (her parents wanted her far away from undue attention). Joel was enrolled in an approved school - the paramilitary kind for wayward boys. It all started with those ice pops!

11.1: Drugs and their legal classification

What is a drug?

A drug is any chemical or substance taken into the body by the mouth (swallowing), through injection or patch on the skin, or inhaling that changes how the body works. A drug may be legal (allowed) or illegal (not allowed by law - a crime).

Legal drugs

Legal drugs can be divided into two categories: over-the-counter drugs and prescription drugs.

Over-the-counter drugs

Figure 11.1: Prescription drugs.

Over-the-counter drugs are medications that you can purchase without written permission from a medical doctor or dentist. These types of drugs are used to treat common illnesses like headaches, muscle pain, or upset stomach. Cold and cough medicine, vitamins, ibuprofen and antacids are examples of over-the-counter drugs. The excess intake/consumption of alcohol is considered drug abuse.

Prescription drugs

Prescription drugs are drugs that can only be given by a pharmacy if a person has written instructions from a medical doctor or dentist. They are usually stronger than over-the-counter drugs and should be used only by the person whose name is on the prescription for a specific illness.

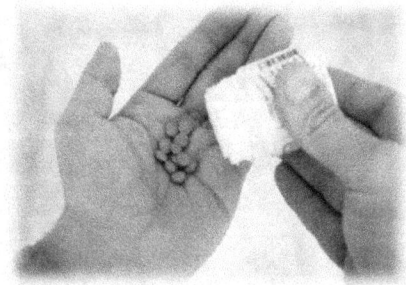

Figure 11.2: Prescription drugs are unique in colour and shape.

For example, if you get a sore throat, you will have to get a prescription for antibiotics to treat it. However, if your brother or sister gets a sore throat, they would need to go to the doctor to get their own prescription specific to their illness. Prescription drugs are used to treat many illnesses, such as heart disease or high blood pressure or problems relating to the kidney, lungs, nerves and skin.

These drugs are safe if they are taken in the proper dosage by the person for whom they were prescribed.

Illegal drugs

Illegal drugs are not prescribed by a doctor and are not permitted by law. There are some drugs such as morphine, heroin, opium, cocaine, lysergic acid (LSD) and those with 'street names' such as PCP or 'angel dust', ecstasy, crack cocaine, Ice and Molly which are illegal and not prescribed. In Jamaica, having more than two ounces of marijuana (ganja) is illegal. For more information on drug use, please visit Home - National Council on Drug Abuse (ncda.org.jm).

Understanding drug labels

It is important to read the product label on medications to find out how to use drugs safely. This is especially important with over-the-counter medications because they are taken without seeing a doctor. Medicine labels provide the following:

- **Active ingredient**: the medicine in the drug
- **Uses**: how often the drug should be used and who should be using it. For example, every six hours and some prescriptions for women should not be used by men.
- **Warnings**: when not to use the product and possible side effects.
    o For example, alcohol must not be taken while a person is taking antibiotics.
    o Some drugs should not be taken if pregnant or intended to be pregnant.

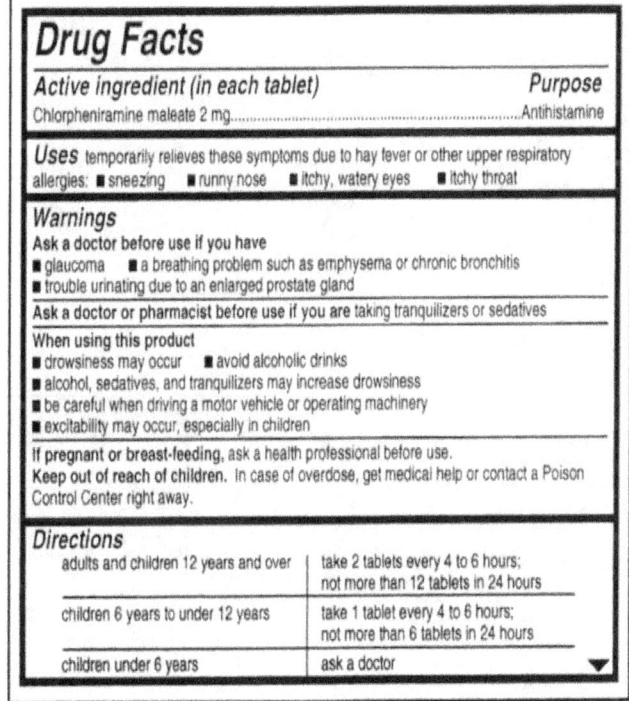

Figure 11.3: *Drug fact label.*

- **Inactive ingredients**: substances like colours and flavours
- **Purpose**: what you can expect the drug to do for you
- **Directions**: how and when to take the medication
- **Side effects**: Possible reaction(s) when taking the drug
- **Expiration date**: when the drug is no longer useful
- **Other safety information**: how to store the medicine

Activity 11.1: Medication survey

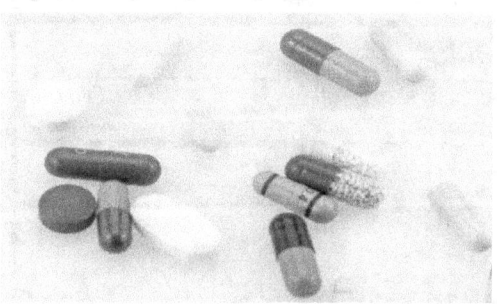

Interview some members of your family. Ask them what medications they take. Ask if the medication is by prescription or if it can be purchased over the counter. Make a table like the one shown to record your findings.

Figure 11.4: Medication

Drug name	Bought by Prescription	Bought without a Prescription	Purpose
1.			
2.			
3.			
4.			
5.			

Let's review prescription and over-the-counter drugs.

1. Define the term drug.
2. Define the terms legal and illegal drugs.
3. When is the use of alcohol considered drug abuse?
4. What is the difference between a prescription and an over-the-counter drug?
5. What three pieces of information can you find on an over-the-counter medicine label?
6. Mary decides to take her mother's prescribed headache tablet. Is this legal or illegal? Explain.

For more information on drug use, please visit **Home - National Council on Drug Abuse (ncda.org.jm)**.

Due to the COVID-19 pandemic, pharmaceutical companies have created new drugs; the vaccines that would help to fight this disease. There are a lot of immunisations (vaccines) that we had to take in the past to prevent measles, chickenpox and other diseases. Yet, many people are not sure whether they should take this vaccine for COVID-19. Why do you think some people may be hesitant?

11.2: Detrimental use of drugs

Using medicine safely

Modern medicine has saved millions of lives by preventing and healing diseases. It has also made life easier by relieving pain from problems like toothaches and headaches. However, if a drug is used incorrectly, it can damage your health.

It is important to use the medicine the way your doctor tells you. This will help you stay healthy and prevent harm. If a drug has expired, it should be disposed of properly.

Drug abuse

When a person chooses to use drugs for nonmedical purposes and pleasure, it is called drug abuse. A person who is abusing drugs may:

- begin losing weight rapidly
- fight or quarrel easily
- have trouble concentrating
- have changes in the colour of the eyes
- sleep most of the time, even in class
- not be interested in school; often absent
- steal
- have friends who do drugs (subculture)
- hallucinate
- think of committing suicide.

Figure 11.5: Drug abuse may lead to suicide.

Using drugs incorrectly can cause health, social and mental problems. Improper use can lead to drug overdose and even death.

Did you know that the side effects of alcohol use can include relaxation and euphoria leading to social withdrawal, violent behaviour, cirrhosis of the liver and birth defects?

Illegal drugs use

Illegal drugs are drugs that are not provided through a doctor or pharmacy. In many cases, people use illegal drugs to feel 'high'—a temporary feeling of excitement, calm or happiness (euphoria).

However, illegal drugs are banned by law because they damage your health and cause addiction. Addiction to drugs can cause you to fail in school or drop out of school. Some drug addicts commit crimes for money to buy more drugs. Over time, drug addiction leads to separation from your family and friends, imprisonment, homelessness or death. There are no long-term effects of using drugs.

Figure 11.6: *Use of illegal drugs leads to bad consequences.*

Drug addiction

Both illegal drugs and legal drugs can cause addictions. A drug addiction makes a person unable to stop using a certain drug when they want to. When a drug addict tries to stop using a drug, they may show withdrawal symptoms such as vomiting, sweating, shaking, hallucinations and fearfulness. However, continued drug use leads to cancers, brain damage, damage to bodily organs, madness or death.

Figure 11.7: Man injecting himself with drugs.

Many people who become drug addicts start their journey into addiction in childhood, hanging out with friends. Drugs are ingested through the mouth and nose by inhalation, through the skin by injection or just touching. Some drugs are ingested by swallowing alone or by mixing in food, juice or alcoholic beverages.

Many young people first experience drugs at nightclubs, dance halls and parties. They get access to drugs through their friends. These friends make sure they get their first experience with drugs free of cost. Before they know it, they are hooked, on the street at midnight searching for a 'fix' (more drugs) and this is when the real problems begin. Sometimes these friends are working with drug dealers and are just other customers.

Figure 11.8: Use of illegal drugs.

For more information on drug use or abuse, please visit Home - National Council on Drug Abuse (ncda.org.jm).

CHEETAH® PREPARING THE JAMAICAN SCIENTIST

Did you know that you can look at a person and know they are on the drug ganja? Some of the physical changes to look for are bloodshot eyes, poor coordination, enhanced sensations and perceptions, increased appetite, dry mouth, possible dizziness and nausea.

Fact Sheet - Marijuana - National Council on Drug Abuse (ncda.org.jm)

Let's examine legal and illegal drugs.

1. What is the difference between a legal and illegal drug?
2. What are THREE harmful effects of illegal drugs on a person?
3. An individual has taken illegal drugs and sees elephants flying around. What could be the effect of the drug taken?
4. The student knows that smoking marijuana after school is wrong but cannot stop and feels compelled to do it. What is the real problem?

Activity 11.2.1: Science, technology, engineering, arts and mathematics (STEAM) and me: The effects of illegal drugs

Create a storyboard to show the effects of illegal drugs. Include the following in your presentation:

- the physical effects of using the drug
- how someone may behave under the influence of the drug
- a written paragraph describing the problems a person may experience while under the influence of the drug
- three places a person can go to for help if they are addicted to drugs
- the steps you would take to help a friend recover from drug addiction.

Activity 11.2.2: Preventing drug abuse

There is an increase in drug use in children aged 10 to 12 years. Using the internet, collect pictures and create a poster for an anti-drug campaign that could be displayed in your community. For instance, ask your local librarian if your class can display the posters for a month for public viewing.

Activity 11.2.3: Dramatisation

During one month, a group of 15 students should meet to write a skit about the challenges of drug abuse. They should set out roles and responsibilities for each person in the group, such as:

- director(s)
- writers
- characters (10 to 12 people)
- props

- plots
- settings.

The plots should include causes and effects situation, conflict, conflict resolution and a conclusion. This skit should be filmed and presented to the entire class.

For more information on drug use, please visit Surveys - National Council on Drug Abuse (ncda.org.jm).

Activity 11.3.3: Information and communication technology (ICT): Separation of mixture

Research to find out how the use of illegal drugs is affecting Jamaican students. Use the following local sources to create a ten-minute electronic presentation.

About Us - National Council on Drug Abuse (ncda.org.jm)

The Gleaner (jamaica-gleaner.com)

Jamaican News Online - the Best of Jamaican Newspapers - Jamaica Observer

Jamaica Information Service – The Voice of Jamaica (jis.gov.jm)

Use an app, such as *Animoto* and add music or sounds to the pictures to create a video entitled, 'Sustainable Development - Protect the future of the environment.'

Evaluate yourself!

Use this evaluation grid to check your understanding of the concepts discussed in this chapter. Read each statement below and insert the symbol that best shows how well you feel you understand the concept. Ask a teacher or parent to help you go over any areas that are still unclear, or that you do not feel you have mastered. Be honest!

I got it!

I need to do more work.

I do not get it. I need help.

In this chapter:

	I got it!	I need to do more work.	I do not get it. I need help.
1. I can state the meaning of the term 'drug'.			
2. I can group commonly used drugs as legal or illegal.			
3. I can explain the importance of guidelines on the proper use of a drug.			
4. I can describe the effects of drugs on the body.			
5. I can show responsible behaviour in the use of drugs.			

CHEETAH® PREPARING THE JAMAICAN SCIENTIST

Strive to learn something new every day. Did you enjoy learning new things from this book?

Let us continue to leap for knowledge!

Glossary

This glossary contains the scientific words and their definitions found in this text, listed in alphabetical order.

A

absorbency: the ability of a material to soak up a liquid.

aquatic: (1) a water-based ecosystem; (2) an organism that lives in water

adaptation: adjusting to survive the effects of changes in the environment.

addiction: a condition that causes a person to be unable to stop using a specific substance or doing a specific behaviour.

air pollution: the addition of chemicals and other harmful substances to the air.

amplitude: a measure of the loudness of a sound. the louder the sound the higher the amplitude and the energy of the sound. Hence high sounds may damage the ear drum.

anus: the organ located at the end of the large intestine, attached to the rectum. The anus allows the waste products stored in the rectum to pass out the body.

arteries: the blood vessels that carry oxygen-rich blood from the heart to every part of the body.

artificial light: comes from objects that are made by humans to produce light. Sources include light bulbs, fires, glow sticks or fireworks.

atmosphere: the layer of gases that surrounds the Earth. These layers also protect living things from the sun's radiation.

B

balanced diet: a diet that contains all of the food groups in the right proportions.

beam: a group of light rays that move in the same direction from a light source.

blood: the red liquid material that carries oxygen and nutrients to the cells of organs and removes waste from the body. It is composed of red and white blood cells, platelets and plasma.

blood vessels: the tubes transporting blood throughout the body; the three types are arteries, veins and capillaries.

bones: components of the skeletal system. Working together, the bones provide support and movement, while protecting the body's other organs.

brain: the organ that controls our ability to think, feel and respond to our environment. The brain is located in the skull and is a part of the nervous system.

calories: a measurement of the amount of energy in a specific food.

carbohydrate: a nutrient that provides a quick source of energy for the body; the two main types of carbohydrates are sugars and starches.

circulatory system: the organ system within the human body that is responsible for carrying oxygen, hormones and nutrients to cells. This system also removes waste from cells.

climate: the average weather seen in an area over a long period, usually 30 years or more.

climate change: a long-term shift in the global climate system which has caused new weather patterns.

colloid: a cloudy solution of suspended particles. These particles do not settle to the bottle of the container that is standing.

colon: one part of the large intestine which absorbs water from digested food and allows the water to pass into the blood.

composition: the components that make up a mixture.

concave lens: a lens that is thicker at the edges than in the middle; bends light rays outward to make objects appear smaller.

condensation: occurs when a gas changes into a liquid by losing heat.

conductivity: the ability of a material to transfer heat or electricity.

constipation: a medical condition in which a person has a difficult time defecating; usually caused by not eating enough fibre.

convex lens: a lens that is thicker in the middle than at the edges; bends light rays toward a central focus point.

D

dairy products: food products that contain or come from milk such as cheese and yoghurt; help with growing strong bones and act as an important source of vitamins and nutrients.

deforestation: the removal of large numbers of trees from a forest area.

diabetes: a medical condition that causes high blood sugar levels; caused when the body does not produce enough insulin to remove sugar from the blood or the cells do not respond to insulin.

diaphragm: the muscle that separates the lungs from the organs in the lower body.

diet: the combination of food and drinks that a person eats.

diffuse reflection: this occurs when incident rays strike a rough surface; reflected rays move away from the surface in different directions.

digestion: this occurs when the food eaten by an organism is broken down physically and chemically to provide nutrients for the body to use.

drug: any chemical taken into the body that changes how body processes work; may be legal or illegal.

E

echo: the reflection of a sound wave to its original source after striking a hard surface.

ecosystem: a community of organisms that live with and depend on each other and non-living materials for survival.

elasticity: the ability of a material to return to its original shape after being stretched by an outside force.

environment: the living and non-living factors that surround and affect an organism.

environmental conservation: the performing of actions that help to protect the planet and conserve its natural resources.

erosion: this occurs when the topsoil of an area is removed by natural processes or human activity.

evaporation: this occurs when a liquid changes into a gas by gaining heat.

excretion: this occurs when waste products are removed from the body.

excretory system: the organ system that works to remove waste and other harmful substances from the body.

exhalation: this occurs when the air in the lung is pushed or breathed out. This air contains extra carbon dioxide released as waste from the body cells.

extinct : all species of an organism are dead; no longer in existence.

F

fat: a nutrient that is used by the body to store long-term energy and cushion the internal organs from damage.

filtration: the process of separating different-sized solid particles in a mixture using a filter.

food groups: the five major categories of foods that humans eat; these include staples, legumes, vegetables, fruits, fats and oils and food from animals.

freezing: this occurs when a liquid changes into a solid by losing heat.

frequency: how often something occurs.

freshwater ecosystem: an ecosystem that contains mainly fresh (salt-free) water, such as a lake, river or pond.

fruits: fleshy, edible seed structure which is sometimes sweet or sour, such as oranges, bananas and strawberries; helps to prevent constipation, as well as helps hair, skin and nails to grow.

G

grains: cereal such as wheat, maize, rice or rye; helps to provide energy to the brain and muscles, as well as prevent constipation.

H

hardness: the ability of something to resist scratching and pressure.

hazardous waste: waste that could be dangerous to humans or the environment. Some hazardous waste is from households, factories and hospitals.

heart: a muscle about the size of a fist located in the centre of the chest that pumps blood throughout the human body.

I

illegal drugs: drugs that are not lawfully acquired.

inhalation: this occurs when the lungs expand as air is brought into the lungs from outside the body. This air has oxygen which is used by the body.

Insulator: material that prevents or reduces the passage of heat, sound or electricity through its particles.

insulin: the hormone produced in the body that helps to control blood sugar levels.

interdependence: this occurs when two or more living things rely on each other for survival.

irreversible change: a change in which a substance changes into a new material and cannot be changed back into its original form.

J

joints: the point where two bones meet and are controlled by muscles.

K

kidneys: two bean-shaped organs in the lower back that remove waste, specifically urine, from the body.

L

landfill: a location where solid waste is disposed of by burying it; typically found in areas that are away from human populations.

land pollution: the addition of chemicals and other harmful materials to the land.

large intestine: a large tube-like structure that allows the absorption of water and minerals into the body; transports undigested food to the anus for excretion.

legal drugs: drugs that are lawfully obtained, for example, with a prescription at a pharmacy.

lens: a transparent material with two curved surfaces that bend light and form an image.

lifestyle diseases: diseases caused by a person's activities or living conditions.

ligaments: strong elastic tissues that keep joints in the body together.

light: a form of visible energy produced naturally (such as the sun or lightning) or artificially (such as a light bulb).

loudness: volume of sound; measured on a scale from quiet to loud.

luminous object: a man-made or natural object that makes its own light.

lungs: a pair of organs found in the chest that moves air into and out of the body. They are in charge of inhalation and exhalation.

magnetism: the ability of a material to be attracted to a magnet.

magnifying glass: a convex lens used to make close-up objects appear larger.

malnutrition: a medical condition caused by an inadequate diet, such as eating too few calories, too many calories or not enough necessary nutrients.

marine ecosystem: an ecosystem that contains mainly saltwater, such as a salt marsh, mangrove swamp or ocean.

melting: this occurs when a solid changes into a liquid by gaining heat.

microscope: a device that uses different-sized lenses to allow an observer to see very small objects that cannot be seen with the naked eye.

minerals: nutrients used by the body to grow bones, make hormones and control our heartbeat; these minerals include calcium, magnesium and potassium.

mirage: an optical illusion that is created when layers of air with different temperatures cause light rays to bend, forming a false image.

mixture: a combination of two or more substances.

mouth: the opening in an organism through which food is taken in and where digestion begins.

muscles: body tissues that help with functions such as moving the skeleton, pumping blood and digestion.

musculoskeletal system: the combination of the muscular and skeletal systems working together to help with movement and protect the internal organs.

natural light: light from living and non-living things such as the sun, stars, lightning and bioluminescent organisms that can produce their own light,.

nerve: the organ in the nervous system that carries information to control actions and reactions.

nervous system: the organ system that controls all the actions performed by other systems of the body. This includes the brain, spinal cord, and nerves.

noise: any unwanted or unpleasant sound.

noise pollution: this occurs when noises become too loud, annoying or harmful to humans or animals.

non-hazardous waste: any kind of waste that can be disposed of in a regular garbage can.

non-luminous object: an object that does not produce light. Some non-luminous objects reflect light from a luminous object.

nose: the organ which brings air into the lungs; also, the main organ involved with the sense of smell.

nutrition label: a label provided on the foods people eat that gives information on the nutritional value of the food; lists information such as ingredients, calories and serving size.

organ system: a group of organs that work together within the body to perform a specific function; the human body contains eleven organ systems.

overfishing: this occurs when fish are removed from an aquatic ecosystem faster than they are replaced.

overpopulation: this occurs when the number of people living in an area is too large for the space and resources available.

over-the-counter drugs: medications that can be purchased without a prescription from a doctor.

O

obesity: a medical condition in which a person is carrying more body weight than is recommended.

oesophagus: the long, straight, muscular tube that contracts and relaxes to squeeze food from the mouth into the stomach.

opaque object: an object that does not allow any light to pass through it. This will produce a shadow when light shines on it.

organ: a group of tissues that work together within the body to perform a specific function.

P

periscope: a device that uses mirrors to allow an observer to see things that are out of their line of sight; usually used in submarines to allow crew members to see what is happening above the water.

pitch: how low or high a sound seems to an observer or listener.

plane mirror: the most common type of mirror; forms images that are life size and right-side up. The left and right sides of an image appear reversed.

pollutant: any substance that pollutes the Earth.

pollution: the presence of harmful materials which contaminate or damage the environment within the air, water and/or land.

prescription drugs: drugs that are available only with written instructions from a doctor or a dentist to a pharmacist.

properties (of matter): the characteristics that make each substance unique.

protein: a macronutrient that is used by the body to perform functions such as making cells, building muscles or bones and repairing injuries; found in products that come from animals like chicken, beef, pork and fish, as well as sources like nuts and legumes.

ray: a wave of light that travels in a straight path from one location to another.

reflection: this occurs when a ray of light hits an object's surface and bounces off.

refraction: this occurs when a light ray bends, as it changes speed while passing from one medium into another.

reforestation: the act of replacing trees within an area where trees have been removed.

regular reflection: this occurs when light rays strike a smooth surface; reflected rays all move in the same direction away from the surface.

reversible change: when material changes and can be changed back to its original form.

shadow: a dark area or shape formed when light rays are blocked by an opaque object.

slash-and-burn farming: a method of land preparation for cultivation in which trees are cut down and burned to create new fields to plant crops.

small intestine: the tube-shaped organ that connects the stomach to the large intestine; most of the nutrients and vitamins needed by the body are absorbed into the body through this organ.

soil: top layer of land found on the Earth's surface. The soil is composed of organic matter, water, nutrients and gases that help plants grow.

soil conservation: the implementation of protective measures such as reforestation,

contour ploughing, and reduction of fertiliser and pesticide use to prevent soil degradation.

soil degradation: this occurs when soil quality declines due to human activity or by natural processes.

solid waste: any unwanted solid materials that are disposed of as garbage. Some solid wastes are in liquid forms such as oil, paint and sewage.

solution: a mixture that has at least one substance (a solute) dissolved into another substance (a solvent).

sound: this is produced when energy is transferred through a medium by a vibrating object.

stomach: a muscular, bag-like organ located at the end of the oesophagus that digests food into chyme.

strength: the amount of force needed to bend or break a material.

suspension: a mixture of particles which are large enough to be seen with the naked eye and settle to the bottom of the container over time, if left undisturbed.

system: A set of parts that work together to perform a specific task.

tendons: strong cords of connective tissue that connect bones to skeletal muscles.

terrestrial: an ecosystem or organism that is land-based.

topsoil: the top layer of the soil.

trachea (windpipe): a hollow tube that connects the pharynx and larynx to the lungs. The trachea allows air to travel into and out of the body.

translucent object: an object, such as frosted glass or a plastic bag, that only allows some light to pass through it.

transparency: the ability of a material to transmit light easily or to be easily seen through.

transparent object: an object that allows most light rays to pass through it; often made of clear or colourless materials.

type 1 diabetes: diabetes that develops in childhood or adolescence and cannot be prevented; occurs when the body produces little to no insulin.

type 2 diabetes: diabetes that usually develops in adulthood because of obesity and high blood pressure; occurs when the body does not respond well to insulin or is not producing enough insulin.

U

urbanisation: the growth of populations of people within towns and cities (urban communities) where many people live.

V

vegetables: plant products like spinach, zucchini and broccoli that are good sources of vitamins and minerals; help to prevent constipation.

vitamins: nutrients that are required by the body in small quantities to perform functions such as healing injuries, fighting illness and producing energy from food.

W

waste products: harmful substances that build up within the body.

water pollution: the addition of waste products, chemicals or other harmful materials to aquatic ecosystems.

water resistance: the ability of a material to repel water.

weather: the short-term conditions of the atmosphere in a specific location, such as the temperature, wind (speed and direction), humidity or air pressure.

wetland: an area of lowland that is temporarily or permanently covered by water.

Let us know what you think about this book. Remember to apply for our CHEETAH scholarship.

www.ingramcontent.com/pod-product-compliance
Lightning Source LLC
Chambersburg PA
CBHW082200070526
44585CB00020B/2215